S. HRG. 114–577

THE TRANSFORMATIVE IMPACT OF ROBOTS AND AUTOMATION

HEARING

BEFORE THE

JOINT ECONOMIC COMMITTEE CONGRESS OF THE UNITED STATES

ONE HUNDRED FOURTEENTH CONGRESS

SECOND SESSION

MAY 25, 2016

Printed for the use of the Joint Economic Committee

U.S. GOVERNMENT PUBLISHING OFFICE

20–442　　　　　　　　WASHINGTON : 2016

For sale by the Superintendent of Documents, U.S. Government Publishing Office
Internet: bookstore.gpo.gov　Phone: toll free (866) 512–1800; DC area (202) 512–1800
Fax: (202) 512–2104　Mail: Stop IDCC, Washington, DC 20402–0001

JOINT ECONOMIC COMMITTEE

[Created pursuant to Sec. 5(a) of Public Law 304, 79th Congress]

SENATE
DANIEL COATS, Indiana, *Chairman*
MIKE LEE, Utah
TOM COTTON, Arkansas
BEN SASSE, Nebraska
TED CRUZ, Texas
BILL CASSIDY, M.D., Louisiana
AMY KLOBUCHAR, Minnesota
ROBERT P. CASEY, JR., Pennsylvania
MARTIN HEINRICH, New Mexico
GARY C. PETERS, Michigan

HOUSE OF REPRESENTATIVES
PATRICK J. TIBERI, Ohio, *Vice Chairman*
JUSTIN AMASH, Michigan
ERIK PAULSEN, Minnesota
RICHARD L. HANNA, New York
DAVID SCHWEIKERT, Arizona
GLENN GROTHMAN, Wisconsin
CAROLYN B. MALONEY, New York, *Ranking*
JOHN DELANEY, Maryland
ALMA S. ADAMS, PH.D., North Carolina
DONALD S. BEYER, JR., Virginia

VIRAJ M. MIRANI, *Executive Director*
HARRY GURAL, *Democratic Staff Director*

CONTENTS

OPENING STATEMENTS OF MEMBERS

Hon. Daniel Coats, Chairman, a U.S. Senator from Indiana	1
Hon. Carolyn B. Maloney, Ranking Member, a U.S. Representative from New York	3

WITNESS

Dr. Andrew McAfee, Principal Research Scientist, Massachusetts Institute of Technology, Cambridge, MA	5
Mr. Adam Keiper, Fellow and Editor of The New Atlantis, Ethics and Public Policy Center, Washington, DC	7
Dr. Harry Holzer, Professor at the McCourt School of Public Policy, Georgetown University, Washington, DC	9

SUBMISSIONS FOR THE RECORD

Prepared statement of Hon. Daniel Coats, Chairman, a U.S. Senator from Indiana	40
Prepared statement of Hon. Carolyn B. Maloney, Ranking Member, a U.S. Representative from New York	40
Prepared statement of Dr. Andrew McAfee, Principal Research Scientist, Massachusetts Institute of Technology, Cambridge, MA	42
Prepared statement of Mr. Adam Keiper, Fellow and Editor of The New Atlantis, Ethics and Public Policy Center, Washington, DC	51
Prepared statement of Dr. Harry Holzer, Professor at the McCourt School of Public Policy, Georgetown University, Washington, DC	61
Questions for the record and responses:	
Questions for the record for Dr. McAfee submitted by Senator Amy Klobuchar	65
Questions for the record for Dr. Holzer submitted by Senator Amy Klobuchar	65
Questions for the record for Mr. Adam Keiper submitted by Senator Tom Cotton	67

THE TRANSFORMATIVE IMPACT OF ROBOTS AND AUTOMATION

WEDNESDAY, MAY 25, 2016

Congress of the United States,
Joint Economic Committee,
Washington, DC.

The Committee met, pursuant to call, at 2:32 p.m. in Room 106 of the Dirksen Senate Office Building, the Honorable Daniel Coats, Chairman, presiding.

Representatives present: Tiberi, Schweikert, Maloney, Adams, Beyer, Paulsen and Delaney.

Senators present: Coats, Lee, Klobuchar, Casey, Peters, Sasse, and Heinrich.

Staff present: Breann Almos, Ted Boll, Doug Branch, Whitney Daffner, Barry Dexter, Connie Foster, Harry Gural, Colleen Healy, Karin Hope, Matt Kaido, Jason Kanter, Christina King, Yana Mayayeva, A. J. McKeown, Viraj Mirani, Brian Neale, Thomas Nicholas, Brian Phillips, Ken Scudder, and Phoebe Wong.

OPENING STATEMENT OF HON. DANIEL COATS, CHAIRMAN, A U.S. SENATOR FROM INDIANA

Chairman Coats. The Committee will come to order. Today the Committee will examine how robots, automation, and technology are transforming our economy. I would like to thank our witnesses for being here, and I will be introducing the three of you shortly.

But first I would like to draw the Committee's attention to the gavel that I just used to start today's hearing. It looks and functions like a typical gavel, but this gavel (indicating), which we normally use, is crafted out of a block of wood and carved down to—a machine will carve this down into the form that it is. This one (indicating) started as a pile of dust, a compound, a plastic compound. Instead of taking a block of wood and carving down the traditional gavel, we have built this—not me; this has been built through the 3D printing process by the Washington, D.C., Public Library's Fabrication Laboratory, or what they call "The Fab Lab," using 3D printing. An amazing advance in technology. Amazing.

Three-D printing works by heating up raw material, in this case plastic, and the compound from which plastic is made, one small layer at a time until the object is completed.

And rather than needing to mold or carve raw material as we did in the past, we now use a—we put our file into a printer and it creates the item according to the user's specific expectations and specifications.

I also have with me a different 3D printed gavel that we will use to adjourn the meeting. I will have to bang it a little harder. But it was made by students at the Washington Mathematics Science Technology Public Charter High School located here in the District of Columbia. What an exciting new world we live in where objects can be manufactured on demand, and with such ease and specificity.

I would like to thank both institutions for their contributions to today's hearing, which tangibly illustrate the topic we are about to explore. I would also like to thank Senator Lee and his staff for helping the Committee prepare for today's hearing.

Recent technological developments have been pushing the envelope faster and further than was expected even a decade ago, making what was once thought of as a science fiction a reality.

I remember the hassle of getting my children to program our VCR. And now my cable box is capable of recording all my favorite shows without me even asking. And meanwhile, some of my grandchildren are probably saying, "What was a VCR?"

The robotic machines are here. Whether it is vacuuming our carpets or assisting in precise surgeries, robots are helping with and performing almost every task that we can imagine. This has led to a greater abundance of consumer products, and more productive and creative workers.

However, as with the Industrial Revolution and previous revolutions, this new robotic revolution clearly is contributing to pressures arising within our changing labor force. Even before these technological advances, America's workforce was starting to age and businesses were beginning to rely much more on automated labor than physical labor. Robots are expected to hasten this trend as they fill in for humans in both blue- and white-collar jobs.

This picture—which I am going to put up; I don't know where it is; we were going to put up somewhere—shows a modern assembly line and illustrates the prevalence of automation in today's economy. Where workers used to assemble vehicles directly by hand, now they oversee teams of precise robots that can weld and assemble vehicles far more advanced than ever before.

We have a number of assembly plants in Indiana. I have been through each and every one of them over a period of my service, dating back to 1981 in the Congress. I am used to seeing that line filled with dozens of human beings assembling parts to the making of an automobile or a truck, and by hand.

Now all I see is a number of robots doing that same process. Where workers used to assemble directly by hand, they now oversee through teams of precise robots that can weld and assemble vehicles far more advanced than ever before.

Automation's rapid progress has also raised challenges with certain government policies. How can we foster an environment where innovators thrive and grow?

How can we foster a social safety net prepared for 21st century labor markets? Do some government policies make human workers prohibitively expensive for employers? How will current workers adapt? And is our education system preparing our youngest citizens for the future economy?

These are important questions, and for guidance we look forward to hearing the views of our distinguished witnesses.

Today we will hear from Dr. Andrew McAfee, principal research scientist and co-founder of MIT's Institute Initiative on The Digital Economy.

We also welcome Adam Keiper—I think I am pronouncing that correctly, Adam—fellow at the Ethics and Public Policy Center and editor of the quarterly technology publication, The New Atlantis.

Our final witness is Harry Holzer, professor at the McCourt School of Public Policy at Georgetown University, and Senior Fellow in Economic Studies at the Brookings Institution.

My thanks to all of you for providing us with your expertise and giving us a glimpse into the possibilities of the future.

I now would like to recognize Ranking Member Maloney for her opening statement.

[The prepared statement of Chairman Coats appears in the Submissions for the Record on page 40.]

OPENING STATEMENT OF HON. CAROLYN B. MALONEY, RANKING MEMBER, A U.S. REPRESENTATIVE FROM NEW YORK

Representative Maloney. Thank you so much, Chairman Coats, for really calling such an important and interesting and timely hearing.

We are here today to discuss the impact of automation on jobs and the economy and how best to harness the immense power of technological innovation.

The United States has long been a leader in this important area, and basic research funded by the Federal Government has played a key role in driving innovation.

We know that automation can boost productivity, lift aggregate demand, reduce consumer prices, and improve our quality of life. While all of these benefits are apparent in the long run, we also know that in the short run innovation can displace workers, causing severe economic pain to workers whose jobs are automated out of existence, or whose wages are reduced dramatically.

Today's hearing is about the future. And let's face it, automation is a difficult thing to predict. We do not know what is going to happen, and we just don't know how fast it is going to happen, or in which industries, or what will be the exact consequences.

One study finds that nearly half of U.S. jobs are at risk of being lost to automation in the next couple of decades. Other studies show that the impacts of automation will not be as great, or felt so soon.

Throughout history, concerns have been voiced that new technologies would make human labor obsolete. It has not happened. While there have been dramatic shifts in how people have earned their livings, the quantity of jobs has increased and the quality has improved.

Yet there are reasons to believe that this could be different in the future. I would like to add some of my questions to the excellent questions Senator Coats put forth:

How do we equip our workers with the tools and skills needed to adapt to the future changes?

What should we do as policymakers to both advance innovation and the expected productivity benefits on the one hand, while also supporting workers adversely affected by technological changes on the other hand?

And how can we harness this engine of prosperity while making sure that benefits are widely shared?

I really am excited to learn more and to hear the questions and exchange here today with our excellent witnesses. But before I yield back my time, I would like to turn to Senator Peters, a former colleague in the House of Representatives. We miss you. And I would like to yield the balance of my time to him. He is the co-founder of the bipartisan Senate Smart Transportation Caucus.

Senator Peters has a deep interest and knowledge of automation and its impacts in Michigan and the rest of the United States, and I yield him the remainder of my time, and it is always good to see you again.

[The prepared statement of Representative Maloney appears in the Submissions for the Record on page 40.]

Senator Peters. It is good to see you, as well, Ranking Member Maloney, and thank you for yielding your time.

As a Senator representing Michigan, I am acutely aware of the incredible opportunities and challenges that automation brings to our economy.

Today the American auto industry is generating connected and automated vehicle technology and mobility solutions that surpass really all of the innovations in that industry's history.

These disruptions will really redefine transportation in the United States and will result in thousands of lives being saved. It will reduce personal insurance costs. It will reduce congestion, and provides benefits to the environment. And these advancements are not decades away.

In fact, in the Model Year 2017 Cadillac CTS will leave the factory equipped with vehicle-to-vehicle technology onboard which NTSA predicts at full penetration could reduce the number of accidents on our roads by nearly 80 percent. And at a time when nearly 40,000 people die on our highways every year, that is a big deal.

As the industry moves towards a world where we have fully autonomous driverless cars that are talking to each other, and to infrastructure, we as policymakers have to start thinking about how to eliminate some of the potential barriers to these developments.

As Ranking Member Maloney mentioned, I founded the Smart Transportation Caucus with my colleague, Senator Cory Gardner so that we can have these discussions about automotive cybersecurity, the future of liability, and other serious implications for the future.

But I am pleased that here today we are talking about what these new technologies will mean for the American workforce when the livelihoods of so many men and women in this country actually depend on the driving of a vehicle, whether it is a car or a truck.

The future of mobility, innovation, and automation presents both great opportunities as well as great challenges, and I look forward to hearing from the witnesses.

And thank you, Chairman, for holding this very important hearing.

Chairman Coats. Thank you, Senator. And thank you, Ranking Member Maloney.

Let me now introduce our panel of witnesses. Andrew McAfee is the principal research scientist at the Massachusetts Institute of Technology, studying how digital technologies are changing business, the economy, and society. In 2014 he co-authored a book entitled "The Second Machine Age: Work, Progress, and Prosperity In Time of Brilliant Technologies." His work has been published in the Harvard Business Review, the Economist, The Wall Street Journal, and The New York Times. He holds a Bachelor's Degree in Mechanical Engineering and a Masters in Management, and a Doctorate from Harvard Business School. We welcome you, Dr. McAfee.

Adam Keiper is a Fellow at the Ethics and Public Policy Center and the editor of The New Atlantis, a quarterly journal about the ethical, political, social, and policy implications of modern science and technology. He has worked on Capitol Hill and various think tanks over his career, and he writes on science and technology policy.

And Harry Holzer is a Professor of Public Policy at the McCourt School at Georgetown University. He is currently an Institute Fellow at the American Institutes for Research, a nonresident Senior Fellow at the Brookings, a Senior Fellow at the Urban Institute, and a Research Affiliate of The Institute of Research on Poverty at the University of Wisconsin at Madison. Prior to coming to Georgetown, Professor Holzer served as Chief Economist for the U.S. Department of Labor and Professor of Economics at Michigan State University. He received his B.A. and Ph.D. in Economics from Harvard.

Welcome, Dr. Holzer.

With that, let me start with our witnesses, and, Dr. McAfee, you can be first and give us a summary of your remarks. And then we will go down the line, and then turn it over to some other Members and work through the question process.

Dr. McAfee.

STATEMENT OF DR. ANDREW McAFEE, PRINCIPAL RESEARCH SCIENTIST, MASSACHUSETTS INSTITUTE OF TECHNOLOGY, CAMBRIDGE, MA

Dr. McAfee. Chairman Coats, I would like to thank you, Vice Chair Tiberi, and Ranking Member Maloney, and other Members of the Committee, for having me here today. It is a great honor.

I want to make four points.

The first one is that the American workforce is very clearly going through some fairly major changes. And to illustrate that point, I would like to show a graph of the post-war United States economy that has four lines on it.

Two of those lines relate to output. They are GDP per capita and productivity over decades of time. And two of those lines are about the workforce. They are about raw job creation. And then median household income, on average. Are we creating good jobs, or not?

And what you notice with that picture is that for several decades, after the end of World War II, those four lines were all going up. That's the direction that we want. And they were all going up just about in lockstep. And then more recently we noticed what my co-author and I call "the great de-coupling." The two lines related to output have continued to go on a pretty healthy upward trajectory, while the two lines related to the workforce have in a sense stalled out. And, by some measures, the median American household or family is worse off in income terms than they were at the turn of the century.

So something is pretty clearly going on.

The second point that I would like to make is that this is a really complicated phenomenon, but one of the forces driving these changes is technological progress. And the way that has been happening so far is that technology has been really good at automating routine work. And by that, I mean both physical work—this of an assembly line in a factory; and knowledge work. Routine knowledge work is a payroll clerk in that same factory. We have had technologies for decades now that have been pretty good at automating that kind of work.

And if I could show my next picture, that is my favorite picture of what happens as technology does its work over time. This is a graph of total U.S. manufacturing output, again over almost the entire post-war history. That is the blue line. And we continue to be a manufacturing powerhouse around the world, and manufacturing output goes up almost every nonrecession year.

The red line is total U.S. manufacturing employment. And that has been on a fairly steady downward trajectory. So this graph clearly shows that we are doing more and more with fewer and fewer workers over time in this industry. It is a trajectory that we are starting to see in other industries, as well.

The third point that I would like to make is, as we are fond of saying in Indiana, we ain't seen nothing yet. And, Senator Coats, the gavel that you showed as an illustration of some of these amazing developments in additive manufacturing, or 3D printing, when I look around at the technology landscape and I see artificial intelligence systems, and deep learning, and machine learning that can beat humans at the games that they themselves devised, when I see autonomous cars and trucks, when I see drones that can move in a swarm and accomplish work together with no oversight whatsoever, I see all these forces coming together.

And the main thing that I think is going to happen is that these phenomena that we have already seen in the workforce, this hollowing out of the middle class, the pressures that we see on the average American family, who that middle class was built on the back of routine physical and knowledge work, I expect these phenomena to continue, and for some of these challenges to accentuate because technological development is not slowing down.

I believe it is speeding up. And it is eating into areas where it has not been present before. It used to be the case that if you wanted to listen to a person and respond to what they wanted, you had to have a human being involved in that work. It is just not the case anymore.

The final point that I want to make, though, is that this is not the time for alarmism and for thinking about—for planning for an economy that has no more jobs. That is just not where we are yet. We are generating on the average of more than 150,000 jobs every month in the country. So we are not yet at the point of peak jobs or peak labor.

Instead, I think we need to kind of retool, or reconfigure some things that we are doing to meet the challenges of this age that we are heading into. And to keep in mind for myself what the right changes, or right policy interventions are, I just keep humming the Old McDonald Theme Song to myself. Because ee-I-ee-I-oh tells me a great deal about where we need to make some changes.

And for me that means education. It means immigration reform. It means facilitating and encouraging more entrepreneurship. It means doubling down on our infrastructure, which is in fairly unhealthy shape. And then finally, the "oh" for me is original research. It is pretty clear that companies are great at applied research, and they tend to under-invest in the very fundamental developments that eventually yield things like the Internet and the iPhone to us.

Thanks very much.

[The prepared statement of Dr. McAfee appears in the Submissions for the Record on page 42.]

Chairman Coats. Thank you, Doctor.

Mr. Keiper.

STATEMENT OF MR. ADAM KEIPER, FELLOW AND EDITOR OF THE NEW ATLANTIS, ETHICS AND PUBLIC POLICY CENTER, WASHINGTON, DC

Mr. Keiper. Mr. Chairman, Ranking Member Maloney, and members of the Committee, thank you for the opportunity to participate in this important hearing on robotics and automation.

In the years ahead, these aspects of technology may profoundly reshape our economic and social lives. A good place to start discussions of this sort is with a few words of gratitude and humility. Gratitude, that is, for the many wonders that automation, robotics, and artificial intelligence have already made possible. They have made existing goods and services cheaper, and helped us to create new kinds of goods and services, contributing to our prosperity and our material wellbeing.

And humility because of how poor is our ability to peer into the future. There is reason to believe that major breakthroughs in automation and robotics are right around the corner, but we should recall that just because we can imagine something does not mean it is actually possible; even if it is possible, that doesn't mean it will really happen. Even if it really does happen, that doesn't mean it will happen in quite the way we imagined it; and even if it does come to pass in something like the way we imagined, there are likely to be all manner of unintended and unexpected consequences.

That said, what do we know? And what do we believe is coming? There are two reasons today's concerns about automation are fundamentally different from what came before.

First, the kinds of thinking that our machines are capable of doing are changing, so that it is becoming possible to hand off to our machines ever more of our cognitive work.

Second, we are also creating new kinds of machines that can navigate and move about in and manipulate the physical world.

The recent blizzard of technical breakthroughs in movement, sensing, control and, to a lesser extent, power are bringing us for the first time into a world of autonomous mobile entities that are neither human nor animal.

To simplify—maybe over-simplify—a vast technical and economic literature, there are basically three scenarios for what the next several decades hold in automation, robotics, and AI.

In the first scenario, automation and artificial intelligence will continue to advance, but at a pace sufficiently slow that society and the economy can gradually absorb the changes. The job market will evolve, but in something like the way it has changed over the last half century. Some kinds of jobs will disappear, but new kinds of jobs will be created. And in many cases we will find new ways for human beings to use and to work alongside machines.

In the second scenario, automation, robotics, and artificial intelligence will advance very rapidly. They will take off. In this scenario, there may be great productivity and enormous economic growth, but jobs may disappear at a pace that will make it difficult for the workforce to adapt without pain. Pressures on American workers in mid-skill jobs will be exacerbated, and there will be new pressures on workers in high-skilled and low-skilled jobs. This scenario could involve severe economic disruption, and perhaps social unrest and calls for political reform.

In the third scenario, advances in these fields will produce something utterly new, maybe something dangerous. This is more of the sci-fi notion you've probably heard about, the "singularity," "superintelligence," things like that. These are strange and radical possibilities, and it's difficult to say much about what they might mean at a human scale.

Now a handful of policy ideas have been proposed that would seek to let us enjoy the fruits of these technological advances while avoiding some of the worst possible effects of disruption.

Some of the ideas involve adapting workers to the new economy. We hear that workers must engage in life-long learning, and upskilling, and they must be as flexible as possible. Of course education and flexibility are very good things; they can make us resilient in the face of what economists call creative destruction. Yet we have to be careful not to place too much of our hope in flexibility since workers are not just workers. They are also members of families, and members of communities. Flexibility can be easier to talk about than to do.

Another proposal one often hears discussed is a universal basic income guaranteed to every individual, even if he or she does not work. This idea has both critics and supporters across the political spectrum. It would present a profound transformation of our economic system but, some would argue, maybe a necessary one if we see a profound shift in the nature of work.

Mr. Chairman, the rise of automation, robotics, and AI raises many questions that extend far beyond the matters of economics and employment we are discussing today—including many legal, practical, regulatory, and moral matters, maybe even existential matters. And I mention a few of these in my written testimony.

I just want to end by saying another word or two about the meaning of work. The science fiction author Arthur C. Clarke said, some four-and-a-half decades ago, that we shouldn't worry about people losing their jobs because of automation. We should look forward to it. We should embrace it. "The goal of the future," he said should be full unemployment. That should be our goal.

That notion raises deep questions about who and what we are as human beings, and the ways in which we find purpose in our lives. Work is not just a matter of toil, but a source of structure, meaning, friendship, fulfillment. In the years ahead as we contemplate the blessings and the burdens of these new technologies, my hope is that we will strive, whenever possible, to exercise human responsibility, to protect human dignity, and to use our creations for the improvement of truly human flourishing. Thank you.

[The prepared statement of Mr. Keiper appears in the Submissions for the Record on page 51.]

Chairman Coats. Doctor, you're on. Thank you.

STATEMENT OF DR. HARRY HOLZER, PROFESSOR AT THE McCOURT SCHOOL OF PUBLIC POLICY, GEORGETOWN UNIVERSITY, WASHINGTON, DC

Dr. Holzer. Thank you very much for having me this afternoon. I would also like to make four points about how technology and automation will affect the labor market, and about appropriate policies to deal with that.

So my first point is that fears of automation and the view that they will eliminate millions of jobs historically have been vastly overblown. We all know about Luddites in Britain. At various times in the U.S. we have had automation scares like in the 1950s and 1960s. This has never, so far, turned out to be true. There has been no aggregate job loss in the long run associated with new technologies, even though individual workers have often been displaced.

But my second point: Even if technology hasn't eliminated large numbers of jobs in the aggregate, it can and has reduced earnings among large groups of workers. In the past 35 years, the digital revolution, globalization, and weakening institutions like labor unions together have reduced employment and good-paying job categories, especially for those workers with only a high school education or less. The jobs most effected were goods-producing jobs for men, clerical jobs for women. But at the same time, wages and jobs have increased for workers who either have the technical skills to deal with the new technology like machinists, technicians, and engineers, or who have other skills that complement the machinery. Those skills could be analytical, or communication skills, or even creative skills. There is a strong skill bias in the technology that actually helps some workers and hurts others.

There is likely a capital bias, as well, that the owners of the capital embodying the new technology profit at the expense of workers overall.

But on the skill bias, Dr. McAfee referred to polarization. We have had growing polarization in the labor market. Growing top. Growing bottom. Shrinkage in the middle. But the middle is not disappearing. And it is not going to disappear any time soon. There is a new middle growing in sectors like health care, IT, advanced manufacturing, parts of the service sector. But those jobs require a lot more education and post-secondary training than many workers in the labor market have.

The problem is that polarization is leading to stagnating or even declining real wages right now for less educated Americans, and especially less educated men. And the declining real wages of less-educated men tend to, number one, reduce their activity in the labor market. A lot of them have simply left the labor market. And also their participation in institutions like marriage. And I think this hurts the overall economy, as well as their families, their children, and their neighborhoods when this occurs. So we want to halt and reverse that wage stagnation.

My third point: Artificial intelligence and robotics are very hard to predict in terms of future trends. It is possible that the breadth and the pace of labor market dislocations will grow, as my colleagues had indicated, but let's be clear. To date there is no evidence whatsoever that this has happened yet.

Productivity growth in the U.S. has actually been declining in the last 10 years. That is exactly the opposite of what you would predict based on all the stories we have heard. The fluidity and dynamism and churning in the labor market have declined in the United States. Again, the opposite of what you might have heard. Now that could turn around. That could change in 5, and 10, or 20 years. Jobs could become more unstable, and they could become harder to find. We just don't see it yet in the numbers.

Which means, number four, future automation should not be an excuse to avoid or eliminate a sensible, moderate set of worker supports and services to address the labor market problems that we have already seen. And several of those problems now exist. Therefore we need solutions on several fronts—the most important being the skill bias of technology.

There is a range of changes we need to make in our skill-producing institutions, especially community colleges, to strengthen workforce services, career counseling, growing partnerships between industry and our skill-producing sectors. Making community colleges more responsive to the labor market with higher accountability is important. I am a supporter of accountability in this sense. Apprenticeships. Career technical education and life-long learning. All of those need to be on the table for improving skills.

I think institutions have to be protected ... not only the right of workers to collectively bargain, which are under assault in various places, I believe we need to support high-road employers who invest in the skills, high performance, and high compensation of their workers. A lot of employers do very well taking the low road, reducing their labor costs at any price. They can do very well in that sense. And that might be what hurts our productivity in the United States.

Thirdly, if the labor market becomes more unstable, we do need to make sure that universal benefits are available and portable. Health care, paid family leave, etc. And then finally, we actually might need to invest in more job creation if the place of displacement picks up and overwhelms the labor market.

So there are lots of issues on the table, lots to discuss, and happy to engage in that conversation afterwards. Thank you.

[The prepared statement of Dr. Holzer appears in the Submissions for the Record on page 61.]

Chairman Coats. Dr. Holzer, thank you. I think this is a fascinating topic here with major implications for the future of the country and for, as you said, individuals, workers, families, our society.

I am going to try to combine a couple of thoughts here into one question and turn to the three of you to respond.

Dr. Holzer, it pretty much goes along the line of what you were saying, because my question was going to be this: We have, if you look back in history, several game changers of immense proportions. You know, moving from an agricultural society to a manufacturing society. And now we are moving into a new type of—is there something different about this phase that we are moving into that separates it from any other?

Can we base some conclusions on conventional wisdom and research relative to what has happened over historical? Or are we seeing something entirely new, that we really cannot totally forecast the direction that it is going?

You raised the question, Dr. Holzer, that I would like to have the others respond to, of the sort of a mystery of why isn't productivity, with all this automation, with machines working 24/7, you don't have to pay for health care, they don't go on vacation, they don't go on holidays, tremendous increases in the productivity, why don't we see that trend in productivity on a much higher trend going up rather than being fairly flat? And participation rate. Is the fact that automation is taking over jobs contributing significantly to our low, relatively low participation rate? And what impact is that having on the participation rate?

So if I could start with Dr. McAfee, this is my question, and ask each of the three of you. And, Dr. Holzer, then you can kind of wrap up there. But what are your thoughts of the other two witnesses regarding this?

Dr. McAfee. Chairman Coats, I think you are asking exactly the right questions, and they are extremely difficult questions.

To your first one, is this time different? The only honest answer is: We don't know. And I agree with Professor Holzer, the historical pattern would lead you to be kind of calm about what is happening. Because we've faced big disruptions before, and our economy grew, our labor force grew, and the American worker was better and better off.

The reason I showed that first picture was to show that there's something that looks fairly different in the data. When we look at job growth, it has tapered off. When we look at average incomes, they have been slowly declining or holding steady for a very long period of time, well before the Great Recession. So there seems to be something new in the data there.

And when I look around at the kinds of technological advances that we're seeing, I try not to get too starry-eyed about them, but they do feel like something new under the sun to me. My way of thinking about it is, most of our previous technologies could only— only encroached a small amount into the total bundle of things that a worker might go to try to offer an employer. So we had technologies that could lift more than we could, that could travel across distances faster, and that could do arithmetic better than we could.

Okay, we bring a lot more to the table than that. We can deal with ambiguous situations. We can understand human speech. We can recognize very, very subtle patterns. That's all fantastic.

I've seen technologies that can do all of those things, do them at a very high level, and I think are either already or very quickly going to achieve superhuman performance in a lot of these areas.

To give one example, if a piece of technology is not already the world's best medical diagnostician, I think it will be fairly quickly. So my version—and again, we have to be very humble and cautious about this—my take is that something actually is different now.

Your other excellent question, why hasn't this shown up in the productivity statistics? There's a huge debate about that. I think two things are going on. Number one, a lot of these science fiction advances that I've been talking about are very, very new. They're honestly, most of them, within the past five years. They just haven't had a chance to defuse throughout the economy very, very broadly yet.

The other thing is, I like to keep in mind an image of two economies. There's kind of an extraordinarily productive, automated, technologically sophisticated one. You know, think of Google and Apple as exemplars of that. And then think of another economy that's very, very labor intensive and only grows by adding more people to the mix. Think of the home health aide as an example of that.

Basically most of the jobs we are adding are in that second low-productivity economy, more so than was in the past. When that is the case, we are going to observe very low productivity growth, as we're measuring it, even though that first economy is ticking along at a very healthy clip.

Chairman Coats. Mr. Keiper, anything you'd like to add to that?

Mr. Keiper. I think that was a wonderful answer. I would like to associate myself with most of that answer. I would just add a point or two as to why there has not been more productivity over the course of the last decade.

The Great Recession is a major cause, or a major explanation that I think most of us would turn to. And just to also amplify what Professor McAfee was saying, it takes time for some of these technological advancements to be picked up and adopted by firms in less high-tech sectors.

So you may see over the course of the next years and decades ahead real advances in productivity in firms that are in, forgive me, stodgier fields than you might see in Silicon Valley.

Chairman Coats. Thank you. And, Dr. Holzer, just to wrap up here, give us your thoughts.

Dr. Holzer. Well I think Dr. McAfee did cover most of the relevant things. I'll just add a few things. So there could be a time lag, as everybody suggests. Robert Solow, the great economic analyst of technical change famously said at the end of the 1980s, if there's all this technological change going on, why don't we see any of it in the productivity numbers? And of course in the 1990s you did see it, briefly.

A second possibility, though, is that the nature of the changes will not be as dramatic as some of those we've seen today. And the biggest proponent of this view is Robert Gordon at Northwestern, who has written a very challenging book where he's saying this stuff doesn't compare at all to the changes in the late 19th century, early 20th century: electricity, the internal combustion engine, indoor plumbing that dramatically changed American businesses and American homes.

So he's a techno-pessimist. He just doesn't think that these technologies will be as ground breaking. But it also could be, as Dr. McAfee said, that a lot of the growth right now is for services that right now can only be done by humans.

A lot of the elder care, child care kinds of work. It is not necessarily very high skilled, but robots can't do that and won't be able to do that. The human touch will not be there for a long, long time. So it could be a mix of these things. We won't know for awhile. But you asked what about the declining labor force participation, and is it directly that the machines are displacing workers and kicking them out of the workforce? I don't think that's it. I think there is an intermediate step having to do with wages.

All the forces that have reduced the wages of less-educated workers—technology, globalization, weakening institutions—as those wages have stagnated and declined, a lot of workers simply don't believe it is in their interest or worth their while to stay in the labor force. And I think that is why they leave, and that is why dealing with stagnant wages through, or separately from the technology, I think is our prime concern right now.

Chairman Coats. Thank you. Thank you, witnesses, for those answers.

Vice Chair Maloney.

Representative Maloney. Thank you so much.

Dr. McAfee, you wrote in a recent piece in The Financial Times that the skills many people have are becoming less valuable in the labor market because of globalization, and also technological advances.

You wrote, and I'm quoting here, quote "We need to figure out how to deal with this situation. This will be one of the most important policy arenas over the coming decade." End quote. And on page 6 of your written testimony today, you note that the Econ 101 Playbook is very clear on what has to be done, but that it is not being followed. Could you clarify what you mean by that? And what could we be doing in a better way in the policy arena to both boost productivity and ensure that more people can benefit from these new technologies? And I would like to hear Dr. Holzer and Mr. Keiper's take on this question, too. Thank you.

Dr. McAfee. Yes. Thank you. Let me try to address the education first, and then broaden out to the whole Econ 101 Playbook.

As one of my colleagues says, the way we are educating people right now—in other words, I believe it is still dominated by rote learning, by the memorization of large amounts of facts and the ability to regurgitate them, and the ability to do fairly basic math and arithmetic, for example. I have a colleague who says those are exactly the skills you need if you're on top of a mountain with no Internet access.

[Laughter.]

Or they are exactly what we needed workers to do 50 years ago. That is just not what we need anymore. And if it is true that technology has been eating into that routine work, we need to be teaching our students and our workers to be good at the less routine stuff. What is in that category? I would say creativity. I would say different flavors of social skills such as negotiation, motivation, persuasion, coordination. There is research that shows how valuable these skills still are. I didn't learn a lot of those in school.

And then finally, just the ability to recognize a problem and to go after that problem, which is a combination I believe of creativity and grit. We are learning a bit, I believe, about how to teach those in schools. Our educational system at the primary level is still dominated by rote learning and fairly basic quantitative skills. And I think that needs to change. I hope it happens quite quickly.

Your broader question that I heard about the Econ 101 Playbook, for example, our infrastructure gets a grade of D+ from the Society of Civil Engineers in America. I just don't see any reason why that needs to be the case.

The entrepreneurship in America, as Professor Holzer says, is on a steady decline. Most measures of business dynamism are actually heading in the wrong direction for the past decade or so.

Figuring out why that is and reversing it I think is critically important. When I think about the fact that a shampooer in Tennessee needs 70 days of training before they can start their job, it makes my head spin. So there does appear to be a thicket of regulations and other barriers that are getting in the way of entrepreneurship and job creation and dynamism. And so dealing with that seems very important to me, too.

And then finally, when you go look around Silicon Valley and the other centers of great dynamism, I am just unbelievably impressed by the number of foreign-born founders and workers and extraordinary contributors out there.

So liberalizing our immigration policies and getting those people who want to build their lives and careers into America seems to me an incredibly straightforward thing to do. I don't know an economist who disagrees. And yet I find our immigration policies kind of Kafkaesk.

Representative Maloney. Dr. Holzer, I would like to bring you into this. What are your comments on it?

Dr. Holzer. We all agree that there's a skills' problem in America, and skills exist——

Chairman Coats. Doctor, push your [microphone] button.

Dr. Holzer. There's many different kinds of skills, different dimensions of skills. Dr. McAfee talked about important general skills—reasoning ability, communicative skills, etc. Those are clearly very important. There are more specific technical skills that matter, as well, that many of our industries like advanced manufacturing are having trouble finding people to do that kind of work.

So we have skill gaps on different dimensions. A big problem is our K through 12 system. Our K-12 system doesn't prepare a lot of workers for the kinds of technical training that often needs to happen at community colleges. So that is an issue.

But the whole institution of community college to me is a funny hybrid. Historically it was a liberal arts institution, a stepping stone to the four-year schools. And a lot of the people there still think that way. Most of the instructors there want to teach anthropology, not machine tooling, or not phlebotomy, and that is an issue.

But the institution itself doesn't respond very well to labor market forces. Institutions get the same public subsidy no matter what their outcomes are. So their incentives to really emphasize career development, career-building skills, I think are limited.

And in general the training for a lot of the jobs that are hard to fill is actually pretty expensive. Equipment, putting equipment on college campuses, is really expensive. Instructors in many of these areas, everything from nursing to machining, are also expensive.

So a lot of the forces are not aligned well for an intermediate institution like community colleges to play the kind of role it can. We all talk a lot about apprenticeships, work-based learning. There are other issues there about why small- and medium-sized employers either don't have the knowledge about that or the resources to do that effectively.

So I think on a lot of different dimensions we could do a lot better. But it is going to take a lot of work, and it is going to take some resources, as well.

Representative Maloney. Mr. Keiper, I loved your quote at the end of your testimony this morning. I would like to get a feel of how extensive are robots in our world. I know that in medical technology they are using robots quite a bit. And certainly Mr. Peters, Senator Peters, mentioned in the building of cars. But what percentage, how much of it is in America? And I welcome anyone to comment on it.

I went to China several years ago and I wanted to see their solar plants, where I expected to see a lot of people running around and working very hard. And what I saw were a bunch of robots walking around with humans managing them on a computer. You could have blown me away. I didn't know that that type of work was there.

So where do we stand? Are we even with the world going forward with robots? Or are others producing more robots than we are? Where do we stand in the globalization of it? How big is it? Is it helping? And then of course how do we protect our workers in it?

Mr. Keiper. All wonderful questions. And some of them are very complicated. I guess it is difficult in some ways to know how many robots there are because different people define them different ways.

You know, if the word "robot" had been around 100 years ago—it hadn't yet been coined—we probably would talk about early household appliances like the dishwasher as a robot. But because it preceded, predated the coining of the word, we didn't apply it that way. And, you know, I think there will be a period when we talk about robotic cars, as driverless vehicles, and then after awhile we'll just start calling them cars.

So it depends partly on terminology. But to answer what I take you to really be getting at, robots are being adopted in developed countries around the world, chiefly in the manufacturing sector. The United States is by most measures leading the world in this area, although other developed nations are doing very well. Germany. China, in some respects. It depends on how you measure: in raw numbers of robots, or per capita.

I just want to say two small additional points, just to piggyback on things that each of the other panelists mentioned. First, something Professor Holzer mentioned about the era of incredible invention that we saw from the mid-19th century through the early decades of the 20th century when we saw electrification, and lighting, and audio recording, and all these amazing things. And by some comparisons, they got the low-hanging fruit, and arguably it's harder to see where the next steps are.

It reminds me of the famous line from venture capitalist Peter Thiel. He says, "I was promised flying cars and all I got is these 140 characters." A reference to Twitter.

[Laughter.]

The point I think really is that—and it's a point Thiel himself has made—is that we need to encourage creativity and looking for major kinds of innovations. It is wonderful that we are talking about driverless cars. I wonder if we couldn't think about greater kinds of innovations and get back some of the optimism that we had in the middle of the 20th century that in some ways seems to have waned.

And then finally, Professor McAfee mentioned the possibility of educators teaching creativity and grit. Can you really teach creativity and grit? I don't know. Maybe. Maybe not. I think teachers can certainly harm creativity and can harm grit. And we have to at the very least do our best to prevent that from happening.

The last thing I'll say, I read recently a biography of Gordon Moore. He's the man who gave us Moore's Law, co-founder and president of Intel. And, you know, really a creative man. And he chalks up some of his success to his kind of wild, outdoors childhood and his early love of chemistry where he was free to just go around in his backyard and blow stuff up. I mean, he just blew stuff up for fun. You can't really do that today. We live in a different kind of world, a different society. And a kid who is just blowing stuff up for fun is more likely to get into some serious legal trouble than to grow up to become one of the wealthiest and most important innovators of the century.

So it is not just a matter of instilling in children creativity and grit, it's a matter of thinking about as parents, as communities, how to not kill those things off when they arise naturally.

Chairman Coats. Thank you, Vice Chairman Tiberi.

Vice Chair Tiberi. Thank you, Chairman Coats, and thanks for holding this fascinating hearing today. Great testimony from all three of you.

Dr. McAfee, I was quite interested in your written testimony when you comment about trade. In my home State of Ohio opponents of trade use—or say, every single job that's been lost is because of trade, not because of technology, or automation, or things that you talk about in your testimony.

You mention in your written testimony that mid-wage, mid-skill jobs like factory jobs—and my dad was a factory worker—are not being created at the same rate as in the past because of globalization and technological progress, which I have seen in Ohio.

The other thing I have heard, as well, is over-regulation. Earlier this year a CEO of a service company told me that because of what he believed was over-regulation by the Obama Administration that some, across industries by the way, were going to use technology sooner than they otherwise would. He gave me an example in the restaurant industry. My first job was McDonald's. That job that I had in the future in America in the restaurant industry probably will not be there for a 16-year-old because technology was going to take that job. And he cited the fact that over-regulation of employment in France—and he had been to France—where now you order in many restaurants not with a person but with a tablet. And the first person you see is when you pick up your food.

And then I heard that a major restaurant chain in the United States just last month, citing over-regulation, was actually testing this pilot out in the United States. So it is that, as well.

So we're not going backward with respect to globalization or technology. The world continues to move ahead, whether we do or not. Whether it is on trade. Whether it is on technology. And as you wrote in your testimony, it has provided unbelievable benefits to the society and to our country and to our citizens.

Yet, as you say, people aren't feeling it. They aren't believing it. Even though they might pay lower prices, and might have better products, they're not associating those benefits with what you talk about in your testimony.

And your testimony highlighted several ways that government can lessen the negative impact of these challenges and side effects. And I was particularly struck by your point about the importance of encouraging entrepreneurship. It helps ensure that we have a vibrant economic system and ecosystem that is constantly regenerating itself.

I have seen it in my central Ohio district, creating employment opportunities for more workers. I am the first in my family to graduate from high school. So I don't know if you've seen this information that was just released earlier this week by the bipartisan Economic Innovation Group. They highlighted some pretty interesting economic terms. The research they released said that new business formation between 2010 and 2014, which the Obama Administration called the time of national economic recovery, that we suffered an unprecedented collapse compared to previous recoveries in our history.

And that, furthermore, new business formation was far more geographically concentrated than in past recoveries in our history. And to put it in perspective, this is an unbelievable statistic, 20 counties in the United States—20—alone, produced half of all net new businesses in the U.S. economy between 2010 and 2014.

So these findings I believe underscore why it is important to support legislation that Congressman Ron Kind worked on together, and that's the Investing In Opportunity Act to increase access to capital in distressed communities, and new enterprises, and startups.

Your testimony highlighted another solution, which is the need to decrease the over-regulation of new innovation. Can you kind of expand on that and elaborate on the policy and regulatory issues that are of most concern to you in relation to entrepreneurship, and stifling entrepreneurship?

Dr. McAfee. Vice Chair Tiberi, thank you for that excellent question because it is an area of increasing concern to me as I look around. A couple of things.

I have already mentioned the fact that there appears to be this increasingly dense thicket of things that an employer or a worker has to confront before they can start something up. And navigating your way through that becomes increasingly difficult, and it looks like more and more people are saying I'm just not going to bother with it.

One of my favorite phrases coming out of Silicon Valley these days is "permission-less innovation." And by that they mean: Let me go do something out there in the world. If it is causing harm, if there are negative consequences, we understand that and we will deal with that. But please don't make me submit my innovation to any kind of oversight committee or bureaucracy, which will then tell me if I can proceed or not.

So I love that phrase. And I've become a big fan of this notion of permission-less innovation. You mentioned the really terrible unemployment situation in France and some other European countries.

It strikes me in my discussions there that they do have a less enthusiastic view of permission-less innovation. There seems to be a greater idea that the state needs to have a role in approving or channeling the course of innovation.

And I certainly don't think that we need to just let innovators go willy nilly and never ever get in their way and let them drive their driverless cars wherever they want to as soon as they want to. That would not be prudent.

But in general, I do think the thicket is dense and getting thicker, and I do worry about attempts or ideas that I hear that take us away from that idea of permission-less innovation.

Chairman Coats. Thank you. Our next Senator, Senator Klobuchar, had I known what I now know, I would have had a robot come in here with a cake in its hand to present to you with candles, light those candles, and we would all sing Happy Birthday.

Senator Klobuchar. Oh, that would have been nice. Thank you. But it is the sentiment that counts. Thank you very much.

I want to thank you and the witnesses for this important hearing. I am really focused on the apprenticeships and preparing our workers for this new economy, coming from a State that already has a 3.8 percent unemployment rate. We just don't have workers to fill some of the technological jobs involved in manufacturing and high-tech work we're doing. And we are going to start losing some businesses in rural Minnesota simply because we don't have the workers.

So that, plus what you all have been talking about with this permission-less—I just like that. It sounds kind of exotic for the Joint Economic Committee—but with the innovation, creates this demand.

And I wondered, I guess I would ask you, Dr. Holzer, about this, I'm just obsessed with doing more with apprenticeships. This country made a MOU with Switzerland, and Switzerland and Germany are doing to get more students into that technological field.

We have a new high school in one of our towns that is amazing. It has tracks you can pick from. One of them is manufacturing. All the equipment is on the first floor. We've got companies that are coming into the schools. We've got community colleges that are partnering because the high schools can't get the machinery but the community colleges maybe can. Just how do you think this should all work?

And what could the Federal Government's role be, when we only fund about 10 percent for education. I figure it's incentives and things like that, but why don't you address it?

Dr. Holzer. Well thank you for the question. If you don't mind, I would use this opportunity to say what we should not do. What we should not do is get rid of all regulation. I was troubled by the tone of the last question.

Senator Klobuchar. Yes.

Dr. Holzer. This implication that regulation is always bad simply is not true. It is not true in theory. It is not true in fact. Minimum wage laws. EEO laws. Occupational Safety and Health. Pension protection. There are a lot of market failures out there. There are a lot of inequities.

If you do regulation badly, if you overdo it, if we had a $15 or $20 minimum wage in this country, I would agree. If the range in which the minimum wage is now are talked about or possibly increases, much more modest, it won't have that effect. So I just thought we needed some balance.

Senator Klobuchar. I appreciate that. I could tell you were getting worked up.

Dr. Holzer. I was having a little trouble.

Senator Klobuchar. Now get worked up about my thing.

[Laughter.]

Dr. Holzer. Let's move on to your question about apprenticeship. There are lots of wonderful examples of things happening out there. In many ways it is hard to scale them. A lot of what we are doing to expand apprenticeship in America, it is often a very retail operation, employer by employer.

Senator Klobuchar. Right. Community——

Dr. Holzer. Providing incentives. And I'm not sure there's any other way to do that, given the——

Senator Klobuchar. Well, Arne Duncan did some grants that were effective in our sort of exurban areas where they would be given the incentive to pair up with——

Dr. Holzer. There are. And we're seeing an expansion of sector partnerships with community colleges. And I like the model. I like where you have apprenticeships with employers where at the same time the worker gets a certificate, or an associates degree. So they get the general skill credentials as well as the specific skills.

But it is simply hard to scale. There are biases in the system. As I said, there's institutional problems. The incentives aren't always aligned. And as I said earlier, if people get to the 11th or 12th grade and they're still reading at the 8th or 9th grade level, you know, that is an inhibitor as well.

Senator Klobuchar. And the other thing is, having been a prosecutor and done all these truancy work, you've got kids dropping out that maybe if you did something while they were in high school, where they were working somewhere, getting a one-year degree, a two-year degree, depending on what it was, that would be good.

Dr. Holzer. I agree.

Senator Klobuchar. You have a whole workforce you are losing. And they can then go on to get more years of school.

Dr. Holzer. That's right. And I really believe that in middle school American students need a lot more exposure to the workforce. Because then they really see that this really boring algebra class actually could be very useful for you in a whole range of well-paying jobs. And then you are more motivated to do it.

And high quality CTE in the high schools. Models like P. Tech and Linked Learning, and career academies, all of those, and there's a pretty solid evidence base on at least some of them, they are all very promising models.

The question is always how do we scale up to replicate those good models.

Senator Klobuchar. Yes.

Dr. Holzer. And also because the German companies are flocking to American, the manufacturing companies. They are astounded at what they see, but they know they can't implement their model. It's got to be an American model and American institutions, and that stuff——

Senator Klobuchar. Um-hmm, and also more women. We were just talking about, Mary Barra, the CEO of General Motors, who was an engineer and worked her way up through the company, and is speaking out on this idea that we need more women, more people of color, because we don't have enough people going into these areas. And for women, especially, the factory floor is no longer dark, dirty, and dangerous, as you've pointed out how high-tech it is.

So I think that is a piece of this, as well.

Mr. Keiper, did you want to respond at all?

Mr. Keiper. Only one additional point. This idea of apprenticeship as it catches on, I think it's wonderful. I would love to see real experimentation in that area across the states. You know, different kinds of models being adopted and tested.

One area that the Obama Administration has said a little bit about, but I think a lot more could be said about, is encouraging our young people to take an interest in the trades. For really what seems like decades in some ways——

Senator Klobuchar. And their parents aren't always encouraging from what they've experienced.

Mr. Keiper. Right. I mean we have encouraged our young people more and more to go into office jobs, and data management, moving paper around, moving pixels around, and I can't help but wonder whether some people, including some of the people who are electing to get out of the workforce, some of the lower-educated men that Professor Holzer was mentioning who are having trouble finding work, whether they might not find other sources of satisfaction if they had been encouraged to look into plumbing, and being an electrician, and roofing, and carpentry——

Senator Klobuchar. And whether they can trade into it now. And we have such a need that it becomes worth having companies finance those.

Mr. Keiper. That's right. We tend to understate, when we nudge our children in those directions, we tend to understate the real value of being your own boss, working in the trades, and we tend to also forget that, you know, of all the jobs that are going to be automated, or going to be replaced with artificial intelligence, you're not going to have a robot doing your roof any time soon. You're not going to have a robot electrician in your house.

The trades, at least for the foreseeable next several decades, are going to require the kinds of problem-solving and dexterity that human beings can uniquely bring to bear.

Senator Klobuchar. That's a nice way of putting it. Thank you so much to all of you.

Mr. Keiper. Happy Birthday.

Senator Klobuchar. Thank you. It was Bob Dylan's birthday yesterday, too. He's from Minnesota. I thought I'd add that in for some glamour.

[Laughter.]

Chairman Coats. Thank you, Senator.

Senator Lee was instrumental in asking for this hearing to be held. I think it is a fascinating subject, and we thank you, Senator Lee, and your staff, for helping us pull this together.

So, you're on. We'll give you a little lenience with time here, for your support.

Senator Lee. Thank you very much, Mr. Chairman.

I would also like to wish Senator Klobuchar a very Happy Birthday. And later on I will be singing Happy Birthday and I'll do my interpretive dance to it, before I'm replaced with a robot.

[Laughter.]

Dr. McAfee, I would like to start with you. Utah has a program that allows elementary schools to partner with certain leaders in the high-tech industry to teach 4th, 5th, and 6th graders basic coding skills over a 16-week period.

That this program exists is exciting. The fact that it is exciting, and the fact that it is somewhat unique and rare ought to be concerning to us.

You know, our schools teach people well, but they generally follow what some might call a 20th century model to educate people and prepare them for what has become a somewhat unique 21st century workplace.

So these educators in Utah are doing great work, innovative work, by helping prepare students for the unique employment skills that they will need in the future.

My question, though, is about what happens next. You know, I think it is possible that what we think of as high-skill jobs today could easily become tomorrow's low-skill jobs, such as in much the same way that the high-skilled welders of yesterday, many of whom have now been replaced at least on assembly lines for automobiles and many other manufactured products, many of them are now without jobs.

So software developers today could find that in many instances their jobs have been replaced through automation. What do you think happens when that occurs? And when we develop technology that does our coding for us?

Dr. McAfee. Senator Lee, it is a great question. And I like your insight that some of the jobs that we consider very safe and very prestigious today might actually flip around because of technological progress.

I think Mr. Keiper got it exactly right. We are not going to automate the work of an electrician, or a plumber, or a prime contractor any time soon. That is still a human job. However, when I look around I see technology able to do a lot of the work that a medical diagnostician does, that many kinds of lawyers do, that financial advisers do.

I recently turned a lot of my assets over to what they call a "Robo Advisor," which generates a mathematically optimal portfolio for you and manages it over time with no human intervention whatsoever.

Senator Lee. What kind of a commission does the robot charge?

Dr. McAfee. Less than a human does. I'll tell you that. That's one of the reasons I did it. So I think there could be this interesting inversion for some professions that we consider, again, very prestigious, very safe right now. They are going to see a lot of automation. Not in the science fiction future, but in the next five to ten years.

Not all of them, however. You mentioned coding. And I think teaching little kids to code is great for two reasons. One is that that job appears to be relatively safe. We have had really lousy luck getting computers to code themselves. Writing good code appears to be closer to writing a novel, and all the attempts at automatic fiction that I've seen are just laughably bad.

So we do need a lot of digital professionals to keep making these technologies for us. More fundamentally, though, I think the reason it is great to teach little kids to code is because it teaches them a style of very clear, very difficult thinking that will serve them very well no matter what they wind up doing with their lives.

One of the efforts that I love in that area is the First Robotics Competition, which has spread like wildfire throughout the country, where teams of kids, teams working together, build an actual robot that competes against other teams' robots.

It was started by a guy named Dean Kamen, who I think is the closest thing we have to an Edison in America these days, who just looked around at this kind of stultifying educational system and thought it was inappropriate for a bunch of reasons. So he has grade school and high school kids build robots and go at it with each other. And it's been a runaway success.

Senator Lee. That sounds exciting. It sounds like a good reality TV program, especially if you end up doing like hand-to-hand combat among robots.

Dr. McAfee. That happens, yes.

Senator Lee. Mr. Keiper, you talk in your testimony about the nature of work, and the value of work both to individuals and to society as a whole.

What do you think the innate human desire to create value and to be valued in society and by society, what do you think is to come of that? And how do you think that will be affected by these changes in technology?

And in particular, as more jobs become automated, do you think for some of the reasons I just mentioned that it would be somewhat hazardous to adopt policies that would involve subsidizing non-work? What would that do to that desire to create value and to lead to new and different jobs?

Mr. Keiper. Well that's a wonderful and very complicated set of questions. Let me deal with the last part first.

We have learned over the course of the last 70 years or so to be very sensitive to incentives and disincentives to work. And, you know, we saw in the major battles over welfare reform in the 1990s huge disputes about how law might incentivize or disincentivize work.

And as some people have begun to talk about universal income, guaranteed basic income as a potential policy solution to a future in which nobody has to work in some distant future because productivity is so high all of our basic needs are taken care of and fewer and fewer people need jobs, that's a really, that's a really complicated idea.

Economists have been fighting about this proposal for decades. Because if you want to create such a thing, you want to structure it in such a way that you're not disincentivizing work. Some of the most interesting and creative approaches to the guaranteed income policy are approaches that structure it in such a way that they encourage good behavior, including involvement in the economy.

I think it's very likely that we'll hear people from both the left and the right really start to discuss and analyze and debate guaranteed income in the years ahead, as that happens it is going to be important to really attend to that incentives question that you're pointing out, the question of making sure we're not disincentivizing work.

Because, as you say, work is—it's not just a matter of doing something that you dislike for which you are compensated. At its worst that's what it can—well, not its worst, it can be even worse than that—but at its minimum, it's that. But it can also be a source of pride, and dignity, self-definition, and meaning and purpose in human life.

And that is sometimes a little—it sounds a little idealistic, I know, but it is really true when you talk to people, and we kind of stop looking just at the economic numbers, but listen to that anecdotes, the "anec-data," as well.

Senator Lee. Thank you very much. Thank you. I see my time has expired. I will save the rest of my questions for the second round. Thank you.

Chairman Coats. Thank you, Senator. Congressman Schweikert.

Representative Schweikert. Thank you, Mr. Chairman.

Can I throw a hypothesis and pitch something at you? Because I have not heard it actually discussed. There are some articles and data sitting here in front of me basically talking about that the entire world has a dwindling workforce. If you actually look at the demographic curve of the entire world, that prime productive age groups is actually on a very steep decline worldwide.

I have other things here basically saying if you add in energy costs, China now is a more expensive place to manufacture than Indonesia, Thailand, Mexico, India. I am looking at many of the charts that actually look at productivity per hour, and many of those countries that we fret about repeatedly here in discussion, actually their labor productive costs per hour are exploding.

Isn't the movement towards automation our way to deal with, first, our domestic crisis here of a decade of—and if you actually dig into the data, it's actually partially, substantially demographically driven, of falling productivity or flat productivity, that if we're going to demand higher wages for our citizens it's going to come with a marriage of talent, labor, and automation, and the fact of the matter is the rest of the world is starting to have a shortage of that productive capacity?

Tell me where I'm wrong. Anyone?

Mr. Keiper. I would just jump in first to say I don't think you are wrong. But two additional points.

One, the causal arrows go both ways. Which is to say, the rise in automation and the increasing complexity of certain kinds of jobs requiring more and more years of education have something to do, according to demographers and economists, with the changing structure of——

Representative Schweikert. Well, but much of that demographic curve is just pure population. I mean, pure age population.

Mr. Keiper. Sure. But age and population I mean are the result of people marrying at certain times, having children——

Representative Schweikert. Higher income, birth rates fall.

Mr. Keiper. Exactly. And those sorts of things are related in complicated ways to the changing nature of work. So that is just a small amplification to what you were saying. Not a point of criticism, but a point of agreement, if anything.

And then I would just add, to kind of further amplify what you're saying, as the demographics change, you look at societies that are aging rapidly, ours is aging not as rapidly as some countries in Europe. Or, you know, we're not entirely sure what China is going to look like. But the one-child policy has had a profound effect on that country's demography.

Representative Schweikert. I'll make your argument. We may have a very good idea.

Mr. Keiper. Well, in some ways we really might. And it's not pretty. And you're going to hear increasingly, and if I'm not mistaken I think Professor McAfee has written a bit about this, and you're going to hear increasingly people talking about the need to use different kind of automation, different kinds of robotics to really work in the growing elder care industry.

Representative Schweikert. Well we are a little off from where I'm wanting to go and I think Dr. McAfee has actually written on a version of this.

My two I want to pitch conceptually is, I have a productivity problem in this country. How do I use the demographics I'm blessed with, which is I'm aging but I'm aging a lot slower than those I compete with, find a way to adopt technology, and that technology may also allow my workforce to work longer?

And I know sometimes that's a whole another side obligation, but if I'm 70 and wish to continue to be in a productive capacity, does technology allow me to maintain my skill sets? And does that help me find a way to continue a productivity curve while the demographic strain on many of those countries I've competed with for such a long time?

I actually think there's optimism if I direct this the right way with the proper incentives. I mean, Dr. McAfee, right or wrong?

Dr. McAfee. Right, Representative Schweikert. I am personally not worried about the productivity stall-out that we are experiencing now. Because when I take the innovations that I've looked around and seen for the past couple of years, and project them forward for 5 or 10 years, our service industry workers today who are not very productive doing things like health care, I believe they're going to become much more productive.

You mentioned this kind of slow ticking demographic time bomb that many countries are confronting. We're confronting a less severe version of it. The reason I think that should worry us is not because we won't have enough people to turn out the goods and services that our economies need. I don't think that's the case at all.

It is because our current workers pay for our current retirees. The social welfare system is configured that way. As these populations gray, I believe that is going to put a lot of strain on these different countries.

Representative Schweikert. Look, I have a fixation on it because I thought it was very well written. In December The Wall Street Journal did that 2050 Series, and some wonderful graphics, some brilliant demography in it, and they make your point. In 13½ years, Social Security is out of money. In 8½ years, Medicare is out of money. And so the underlying theme of these sorts of discussions is: Okay, if you really care about maintaining of these social contracts, how do we dramatically increase productivity so we have that thing called, oh yeah, money.

Thank you, Mr. Chairman. Yield back.

Chairman Coats. Thank you. Senator Peters.

Senator Peters. Thank you, Mr. Chairman. This has been a fascinating hearing. I've got a number of questions, but I thought I would start with something that I mentioned in the opening comments, which is something I have been very involved in being from Michigan and what is happening in the auto industry, and the new, transformative technologies that are coming online. In fact, folks are saying with the new autonomous features will be more disruptive, or at least equal to when the car first came off the assembly line. That is what we are talking about.

But I would like to get each of your perspectives on autonomous vehicles. What's happening in mobility, as you are looking at the greater context of what automation is going to do to jobs. You know, this is certainly a central piece that all of you have mentioned in your testimony or in your writings.

Give me your assessment of that. As policymakers, what should we be thinking about as we are thinking about autonomous vehicles, specifically? We'll start with Dr. McAfee.

Dr. McAfee. Senator Peters, thank you. I called in all of my favors in 2012 and got a ride in one of Google's autonomous cars. And at the time I felt like an astronaut. I felt like I was having an experience that very few other people ever had, or would get to have going forward.

Four years later, I went for a test drive in a Tesla, just a month or so ago, and it has a completely autonomous highway driving feature on it. So in four years that technology went from kind of astronaut-level rare to anyone can walk off the street and experience this.

I think that is a great example of how quickly technology is progressing these days. And to your specific question, I think that completely autonomous vehicles are, if they're not feasible today, they will be very, very quickly on American roads in traffic. Maybe not rush hour Manhattan, but certainly across our interstates.

The main implication of that, I guess there are two: One is that fewer lives will be lost. Safety will go up. The goods that we ship across the country will get a lot cheaper. It will be a great boon for the country in a lot of ways.

The challenge is I believe in a majority of U.S. States today the single most common job is truck driver. And I don't think that will be the case even a decade from now.

So again, preparing for that and dealing with that aspect of this hollowing out I think is going to become an increasingly urgent policy arena for us.

But my prescription is never to try to turn off the autonomous car. It delivers great benefits to us. And I think American auto manufacturers want to be in the lead on that technology.

Senator Peters. Absolutely. Absolutely.

Dr. Holzer.

Dr. Holzer. So I agree with Dr. McAfee that the losers in this will be motor vehicle operators, truck drivers, bus drivers, etc. And that has been an important occupational niche for relatively unskilled workers in America.

Senator Peters. One of the largest in fact, isn't it?

Dr. Holzer. Pardon me?

Senator Peters. It is one of the largest occupations.

Dr. Holzer. That's right. That's right. So that could be a big change. Now it depends. The diffusion of this new technology into use by lots of businesses, I'm not sure of the speed of that transition. It may be very quick. But there may be all kinds of things where the judgment of a driver is still required.

So when I think of UPS and the skills of a UPS driver have actually gotten pretty high in terms of being able to do all kinds of GPS, figuring out the location of the customers, the tracking of the products.

I don't know that driverless autonomous vehicles are going to eliminate all those needs for that kind of work. So I can see that in some sectors of transportation it might be very rapid; in other sectors, more slowly.

But let's suppose it's rapid and we start having this problem of displaced truck drivers, bus drivers, etc. What do we do with them? We haven't talked much about that.

So to approach this, one is to try to teach them a new skill. That's what people mean when they talk about life-long learning, sending people back.

It's a little easier when you're talking about a welder, which is one of the examples before. Old-fashioned welding vs. new precision welding. You have a strong base and you just need a little bit of tweaking.

Our success with retraining and re-educating is going to depend a lot on the age of that bus driver or truck driver. And, number two, on their underlying skill set. Right? So it's easy to send a 30-year-old back to community college than a 50-year-old. It is easier to send somebody back who actually is pretty good at reading manuals and handling technical material than someone who is not. So it is going to depend on that.

For people, there's going to be a lot of people for whom it simply doesn't make sense to retool and retrain. And for those, I frankly think the best thing we can offer them is some kind of wage insurance. Wage insurance does incentivize people to work, not to stay on unemployment insurance. It incentivizes people to shift to a lower-wage job, but then the government, the Federal Government, makes up part of the loss.

So if you have to downshift from a $20-an-hour job to a $10-an-hour job, the government might pay half of that difference for two or three years. Of course we'd have to fund that, and that would require some additional resources. But that is another way of protecting some folks who simply can't be retooled and sent back, but others can. And I think we need to be doing some of both to prepare for that kind of world.

Senator Peters. Thank you.

Mr. Keiper.

Mr. Keiper. Just a quick follow-on to what both of the other panelists have mentioned. If you want to encourage and want to take advantage of the many benefits of driverless vehicles, I would urge you to encourage the legal and insurance industries to continue the work that they've already begun in sorting out questions of liability and damages, which in some ways may present greater hurdles than the technical problems, many of which are already solved.

Senator Peters. Well, my time is about up, but you are absolutely right about that; that sometimes policy is a lot slower than technology, and the technology is moving very, very rapidly. And I would add that, in addition to some of the insurance issues also cybersecurity is a big deal, given the fact that it is bad enough when someone steals the money out of your bank account; it is worse if they drive you into a wall. So those are two other things that we have to do here as policy makers.

Thank you to all of you. Appreciate it.

Chairman Coats. Yes. Fascinating questions. No one has yet raised pilotless planes. I'm not sure how many of us would want to board that plane right now. On the other hand, driverless 18-wheelers gives you pause about what's coming down the road at you. And there's going to have to be some education with the public here and demonstrations, I think. But I assume it is just as easy to put people from location to location on a pilotless plane as it is in a car. It is something to ponder as you lie in bed at night thinking about the future.

Congressman Beyer.

Representative Beyer. Thank you, Mr. Chairman.

And I want to begin by piling on Mr. Keiper's comments earlier about changing the education system. The Senator and Chairman served in Germany and saw it first hand, and I did in Switzerland, where they were taking kids at 13 and 14. In Switzerland, literally two-thirds of the work of the children's force and channeling them into plumbing and roofing and auto mechanics, but also nursing, and radiology, and pharmacy assistants. And so last I checked, the unemployment rate was 1.8 percent. And year after year they turn out enough kids at 19 for the jobs that exist. It is tough when they get to be 50 years old and they're a coal miner, but we have to try anyway.

Mr. Keiper, you wrote about universal basic income and negative income tax, and what do you do when actually a significant part of the population doesn't need to be employed. And you wrote, also you said that it's been discussed favorably for various reasons by prominent conservative and libertarian thinkers.

Can you expand on why they would be for that? I understand the left-leaning progressives——

Mr. Keiper. Sure. Before I get to that, let me just say, yes, I've gone to, visited Germany and Switzerland and studied their education system and taken an interest in it. And while I don't think it would be perfectly applicable here, there may be some states that might want to experiment with adapting American education in something like that direction.

As far as the guaranteed basic income, it's something that Hayek talked about, something that Milton Friedman spoke about favorably, something that I think Charles Murray has written and spoken about favorably if I'm not mistaken. I'd defer to my economist colleagues here to say more about that. But my sense is that folks on the conservative and libertarian side of the spectrum see it as a couple of things.

One, it is a way to continue to encourage innovation.

And, two, it would be a—I think this is Friedman's approach—it would be less complicated than continuing the bureaucratic system of welfare that we had in place when he was writing about it.

He was worried about, you know, all of the costs and pressures caused by this complicated system of welfare that we had in place. He thought maybe if you replaced it with something that was simple and clean and seamless, that that would be an improvement.

There's much more to it than that. There's all kinds of economic modeling that, you know, my colleagues here could say much more about I'm sure, but it's interesting. It's surprising how you get people on the left and the right talking about this with interest. Very little interest in the kind of moderate middle, although I think that is likely to change in the years ahead.

Representative Beyer. Dr. Holzer.

Dr. Holzer. So I'm going to speak as someone from the moderate left-center position who is skeptical. I understand that this could be necessary down the road. I am not anxious to go there anytime soon for a couple of reasons.

Number one, we have a tax revenue problem in this country just funding the liabilities we have in Social Security and Medicare. And we have, politically, an issue—to get the demagogues out—we have a massive resistance to almost any kind of tax increase. So this would require, after we've already paid for Social Security and Medicare and Medicaid, this would be a massive expenditure of tax revenue on top of that. And it's hard for me to see the American public any time in the near future going there.

Representative Beyer. If I can interrupt you for a second, I only have a minute-and-a-half left, I have a very specific thing for you. Most of the new economy jobs have not lent themselves to unionization. How should the labor movement adapt to this new economy?

Dr. Holzer. Not very well, necessarily.

Representative Beyer. Moving forward? I struggle with the Labor Day speeches. What do I say?

Dr. Holzer. With good reason. Some people think that the model of unionism that we have in America was developed mid-20th century. It was very well suited to that kind of economy, and it is less well suited to an economy with a lot of dynamics, and a lot of fluidity in and out of jobs, and dramatic new competition.

So if you look at the SEIU, as one industry, they have found niches. They have been successful unionizing hotel workers in places like Las Vegas and LA that we might not have anticipated. But also they take on more political roles, right? So they work hard for candidates and for policies like a higher minimum wage. I think we're struggling for a new institutional model that fits the new reality of the 21st century labor market. But in the meantime, there are some important roles to be played for that institution. And we've seen actually, the Communications Workers are very good at expanding the skill set of their workers. So I think some of these unions have done a good job of finding a new niche and a new role to play in the 21st century labor market.

Representative Beyer. Okay, thank you, Dr. Holzer. Mr. Chairman, I yield back.

Chairman Coats. Thank you.
Senator Casey.

Senator Casey. Mr. Chairman, thank you very much. I appreciate the testimony today and appreciate the work that goes into your appearance today and what you've told us already in the testimony.

I want to tell the Chairman that as we get an opportunity once in awhile to brag about our states, Pittsburgh, Pennsylvania, in many ways has really been at the forefront of robotics' automation for years. Literally, not just years but for decades. And they are into a brand new chapter.

In fact, there is a neighborhood in Pittsburgh they used to call the so-called "strip district," but now is known also by the phrase "robotics row." Just looking at some of the companies there now, Near Earth Autonomy is one. Real Earth is another company that performs the world's most accurate indoor 3D mapping, for example.

The Near Earth Autonomy operation has an aerial robotic platform to analyze crop data that will help accelerate plant breeding. So on and on with brand new technology. And that is some of the new breakthroughs that we are seeing in a community like Pittsburgh, which had to recover over the last several decades from the collapse of steel and the larger manufacturing economy going in the wrong direction, now that they've figured out a way to invent a new future.

But I was struck by Doctor—Dr. McAfee, when your charts tracking what's happened, the combination of what's happened to workers and productivity at the same time. One chart that I remember from going back about two years ago now was the original, at least what I remember as the original version tracking productivity and wages, similar numbers where World War II to 1973, productivity was up 97, wages were up 91; 1973 forward, productivity still went up over I think it was 72 percent. Wages only went up 9.

So no matter what chart you use, or how you track it, there's been a disconnect between wages and productivity.

So with that as a just kind of a basic foundation for my one and only question, we'll start with Dr. Holzer and anyone else who wants to weigh in:

With these advancements in technology and also with the knowledge we have about what has happened to wages over time, even when productivity was up, wages have been basically flat, how do we ensure that workers get their fair share of the benefits of these technological breakthroughs, the type of breakthroughs we see with automation and robotics and otherwise?

Dr. Holzer. I think first of all that that disconnect in the chart between productivity and earnings is a complicated one. So for instance the rise of health care costs certainly accounts for part of that. That eats up part of it. So if you show average compensation rather than average wages, it is not as much of a disconnect.

If you use mean vs. median, too much earnings have gone to the top 1 percent, sometimes because they're superstars and they deserve it. Sometimes because we have terrible market failures in our financial sector. So there's all kinds of things going on.

The best thing we can do is to educate our young people in skills that complement the new technologies, in skills that are not made obsolete. So playing in a symphony is not going to be outsourced or replaced by a robot. But, frankly, doing good child care and good elder care also is not. Those tend to be low-wage occupations, sometimes lower than they really should be, but there's a range of skills, some technical and some creative, in all of those kinds of occupations and skills and we want to do a better job of educating our young people.

And then on top of the reforms in the system to try to deal with some of those terrible market failures that have caused too much of our income to go to the top. So I think it is that mix. But the best bet remains skills that are adaptable over people's life span, specific to the current job and career, but still where they have some flexibility to retool. So it is a mix of general and specific skills. And again, looking towards the things that complement technology that will not be easily replaced.

Actually in Pittsburgh the Eds & Meds——

Senator Casey. Right.

Dr. Holzer [continuing]. Interaction has been dramatically important. And again, those are the education sector and the health sector. Some diagnostics can be done through technology. Many cannot. Many of the elder care for patients. And so that is an interesting example of ways in which technology will not wipe out everything we care about.

Senator Casey. I know I'm out of time. Mr. Keiper, Dr. McAfee, if you want to answer in writing, or whatever the Chairman would prefer?

Chairman Coats. No, go ahead and answer.

Dr. McAfee. Very quickly, Senator Casey. I think my colleague's answer is excellent. Education and reskilling is obviously crucial. The only thing I would add is, I am a huge fan of an expansion of the Earned Income Tax Credit. Whereas, if we have people who are trying to do the right thing and get out there in the workforce and work at a relatively low wage job like child care, like elder care, let's top their wage up with a subsidy.

I actually prefer that to a substantial increase in the minimum wage because that might have a disincentive effect on the employers from employing that person. I would like to see our system work so that we directly incentivize both the employer to employ that person, and that person to get out there and do that job via topping up their wages.

Senator Casey. Okay. Mr. Keiper, either now or in writing, whatever is best.

Mr. Keiper. I think those were excellent answers. I guess I would only add, if you look further out into the future, some of the theorists who think about these things worry about the possibility of a shift in the balance between labor and capital. And I think that has come up. We have kind of alluded to it a few times here today, which is to say, we can imagine a future where the people who are the first investors and owners of some of these more advanced automated technologies are going to reap much more of the profit—since many machines will bring more value than workers because there's going to be less need for certain kinds of work. That's a possibility. It's very complicated, and there's good reason to think if that does happen it won't last because many of these technologies may be democratized. They may become more affordable for more people.

It's hard to say, though.

Senator Casey. Thank you very much.

Dr. Holzer. It might be an argument for profit sharing.

Senator Casey. Okay. Thank you.

Chairman Coats. Dr. Adams.

Representative Adams. Thank you, Mr. Chairman, and Ranking Member. Thank you for hosting today's hearing. And, gentlemen, thank you for your testimony.

Manufacturing is an important industry in the State of North Carolina that I represent, and it has been impacted by automation. And while automation may increase productivity and improvements can displace manufacturing workers, manufacturing output has partially grown since the Recession, but we still have a ways to go in recovering jobs that were lost, which has been to some extent exacerbated by automation.

But I think part of helping employment in the manufacturing industry rebound is understanding that automation and the skilled workforce can co-exist. And as a matter of fact, I think it is critical that they do co-exist in today's global economy.

Dr. Holzer, we have seen growth in the manufacturing sector in terms of output over the last several decades, with the exception of contractions within the industry that took place during the Recession. In contrast, employment in the manufacturing sector has declined, partially attributable to the growth of automation.

So what role do you think automation played in the previous decline in U.S. manufacturing employment in the early 2000s? And how can we ensure that job growth and automation are working in tandem as our markets continue to evolve technologically? That is for Dr. Holzer.

Dr. Holzer. So there's different kinds of manufacturing. There's high-end advanced manufacturing which we've focused a lot on today. And then there's the apparel sector and some of the nondurable, which are less technologically advanced, but in fact those are relatively low-wage jobs, although important to a state like North Carolina in many cases.

So I think, I think imports from China in the early 2000s devastated nondurable manufacturing. That had nothing to do with the business cycle, just the massive growth of China producing reasonably high quality products for very low wages, and our nondurable sector simply couldn't compete with that.

And that is going to happen sometimes. Nondurable manufacturing is just not a sector where I think Americans can concentrate. And again, they are increasingly becoming lower-wage jobs anyway.

In durable manufacturing, it's a very different story. China has not played nearly as much of a role in durable and advanced manufacturing. And some companies that have actually gone to China sometimes have insourced back into America because—and in fact that's why so many German Companies are here. I know in North Carolina hundreds of manufacturers from Germany have entered North Carolina because of the advantages of low taxes, low regulation, proximity to consumers, all make the United States competitive in a lot of advanced manufacturing industries.

There the big problem is the skills. And do enough American workers—in North Carolina, as you know, when Siemens built their gas turbine engine plant, I think about five or six years ago, they refused to build the plant until they had worked out the skills' problem with some of the UNC, with Piedmont Community College and UNC, which indicates that the skills problem is crucial for the innovation to continue, and it is also crucial for the workers to share in the benefits of that.

So it goes back to the issue of how do we make sure that our young people have those technical skills to be able to be complements with that machinery. It has to start early in the K through 12 system. I think high-quality career and technical education can be a very important part of this. Not old-fashioned voc ed to attract minority kids away from college; high-quality career education. Exposing young people to technical skills and employ building skills much earlier.

We need to make sure that those jobs can be widely shared and the benefits distributed widely. And for folks who can't do that, the Earned Income Tax Credit is extremely important, as is wage insurance, as are very moderate increases in the minimum wage.

Representative Adams. Okay. That was going to be my follow-up question to you. You've answered it.

But let me ask about tax credits and how important you think they are in terms of also workforce training incentives, and increasing productivity and innovation. How are these things going to be important, do you think?

Dr. Holzer. You mean tax credits, for instance, for employers?

Representative Adams. Yes.

Dr. Holzer. Um, we don't have strong evidence on that right now. I think South Carolina is a state which is paying employers a $1,000 tax credit for every new apprentice they take on. I don't know if that's going to work or not. It is an interesting model. What we need here is a lot of experimentation and a lot of evaluation to see what the impacts are.

So tax credits could play an important role to try to help and incentivize. There are a lot of employers who do very well in what we call the low road system. A combination of tax credits and technical assistance might convince more of them to upgrade their skill content in their automation.

Representative Adams. Thank you, sir. My time is up. Mr. Chair, I yield back.

Chairman Coats. I am going to turn to Congressman Beyer again. He would like to start a second round. And then, given the schedule issue I have, I am turning over the gavel made by the students at the Washington Science Tech's Charter School to my colleague, Senator Lee. I know he has some other questions he would like to ask. So I am happy to turn that over to him. He was instrumental in helping to put all this together, so I hand you the 3D designed gavel, and manufactured gavel, as well, if that doesn't work, here's the old-faithful wood carved gavel. And you've got the Chair. Do whatever you want to do.

Such power we're handing over to you.

[Laughter.]

Mr. Beyer.

Representative Beyer. Thank you, Mr. Chairman, and your excellency. I said two questions.

Dr. McAfee, there was a lot of discussion today about the slow rate of increase in labor productivity. And you did mention earlier that that is controversial, all the measurement techniques. And I just look at my life, and the lives of people around me, and email, how much we get done per day.

I mean I think we communicate five or 10 times as much as I did 20 years ago. Twitter, and the fact that we know what happened in the news 30 seconds ago, as long as we stay on top of it. Telework, which is exploding, at least through the Metropolitan Washington Area.

And it's a terrible thing to say as an employer, but people always work harder at home. You know, the 24/7, on Saturdays, and Sunday nights. Or, medicine. The fact that you can get blood tests in 20 minutes, or 15 minutes, or an X–ray a minute later, or—my brother just got a new hip and went home that afternoon. Or even Amazon, where you order the book in the morning and it shows up that afternoon courtesy of the U.S. Postal Service.

How do we make, as an economist how do you think about how we make real progress in measuring what the difference in productivity actually is?

Dr. McAfee. That is an extremely tough question, sir, because as you are pointing out, you gave some wonderful examples. None of them show up in the productivity statistics.

Representative Beyer. Right, right.

Dr. McAfee. They are invisible. But they are clearly increasing our welfare. As you point out, they are increasing our health, which might be the most important thing of all. They are increasing the convenience that we enjoy in our lives, and they are not visible in our classic economic statistics.

These are some of the reasons why I am actually not that bothered about the labor productivity slowdown as we're measuring it, because that was never intended to be a welfare measure. It is not a measure of how well we are doing as an individual or as a society overall.

Some of my colleagues are trying to work on correcting those measurement errors and coming up with more comprehensive ways to think about the betterment that technology brings to us. So stay tuned for those.

I just want to repeat a point that I made earlier, though, which is even with all those problems, I am pretty confident that over the next five to 10 years labor productivity, even as poorly as we're measuring it, is going to go up a great deal because the technologies that I've seen over the past couple of years I believe are going to diffuse pretty rapidly throughout the economy and improve some of these industries like health care.

Representative Beyer. I find just the Cloud, itself. You know, a little family business. We're going to save a million dollars this year by moving from processors in-house to the Clouds.

Dr. McAfee. And that is a really good example, because the Cloud is actually shrinking GDP. The Cloud is actually shrinking some software companies' revenues. As a result, the way we're measuring GDP, and therefore the way we're measuring productivity, will show a decrease. And that was the Cloud making your business and lots of other ones better off. Absolutely it is. It has been a huge, huge advance.

Representative Beyer. Dr. Holzer.

Dr. Holzer. I have a slightly different take on this. And I have disagreed with my two colleagues remarkably little all afternoon, which has been very nice.

This bias that GDP mismeasures, it doesn't measure quality changes, for instance. That bias has always been there. It's not clear that the bias is getting worse.

There was actually a recent paper from the San Francisco Federal Reserve Bank trying to measure that that shows, if anything, that bias has not gotten worse and does not account for the productivity slowdown.

So I am a little less sanguine about that. And I don't think people's earnings—and there's a lot of reasons why, separate from the bias issue, why productivity might not be rising. There's an economist at Princeton named William Baumol. He was the inventor of what's called Baumol's Disease where he said in some parts of the service sector, like the symphony, you can't raise productivity. It doesn't matter how great the robots are. And he sees those sectors expanding, that there are certain kinds of tours that productivity just isn't going to grow, and they've been an expanding part of the economy.

That is a problem to the extent that without that productivity growth wages are going to stagnate as well. The other possibility is that, I am troubled by the gap between profitability and productivity.

It is a conundrum we have. We have very high profitability in the U.S. economy right now. And that's fine. I'm not against profits. But while productivity is stagnant, that's a puzzling contrast.

And one of the arguments is that companies have found a lot of ways to make money using very few workers, outsourcing, offshoring, turning their workers into independent contractors where they're responsible for benefits. There's a lot of that where the incentives are not necessarily to improve productivity and output. It's to minimize labor costs at any price.

So I am a little more troubled by the productivity trend. And maybe it will turn around. Maybe it will be just like the 1990s. Maybe we're at the cusp of another Solow moment when productivity is going to take off. But I just think there's, separate from the measurement bias, other incentive problems holding back productivity growth and therefore earnings' growth.

Representative Beyer. Thank you. Mr. Chairman, might I beg one last question?

Senator Lee [presiding]. Sure.

Representative Beyer. Mr. Keiper, Elon Musk, Stephen Hawking, and others have warned us about emergent properties, consciousness coming out of all the AI stuff. Do you have any concern about that, any worries for us as policymakers? Think Terminator.

[Laughter.]

Mr. Keiper. These are interesting questions, radical possibilities. And at the moment they don't rise to the level of anything that anyone on this Committee ought to be concerned about.

They are things, however, that folks in the technology fields, futurists, and academics should and are thinking about. And I think just for that subject, I would just leave it at that.

Representative Beyer. Thank you very much. Mr. Chairman, I yield back.

Dr. McAfee. Could I add one word to that? Because I get asked this question all the time, and I asked this of the artificial intelligence researchers who I get to interact with, and I heard a great response from one of them. He said:

Worrying about the singularity or the summoning of the demon because of artificial intelligence is like worrying about over-population on Mars.

Representative Beyer. We're working on that, too.
[Laughter.]

Senator Lee. I worry about that a lot. But after reading the book "The Martian" and then seeing the movie, you know, I'm less concerned than I used to be.

I want to thank all of you for being here today. This has been a very informative hearing. As the hearing is evidence, we live in an era of rapid and dramatically transformative technological change.

Within the past 20 years, the Internet has catalyzed a global, social, and economic revolution, one that has affected almost everyone all over the world. We have gone from bag phones in our cars, phones that are big to phones that fit in a bag, phones that are the size of a cinder block to phones that are tiny, in just a few years. And now we have not only phones but super computers in our pockets.

The incredible science that is being used to make self-driving cars a reality and robots that can walk on two legs while maintaining balance, and lift heavy objects, is also creating pretty massive and largely understandable fear of the unknown.

As former Secretary of Defense Donald Rumsfeld once famously said: There are known knowns, known unknowns, and there are also unknown unknowns. He had quite a way with words, still does. It is those unknown unknowns that cause serious uncertainty and sometimes lead to fears about the future.

That is why we held this hearing today, was to explore those unknowns, particularly the unknown unknowns. The value of this technology to the world economy and to our domestic economy is of course enormous. The impact of this technology will go far beyond the bottom lines of businesses, though.

It is and it will continue to have an impact on the nature of work and, by extension, the ways in which we live and define our lives. But it will also impact our society more broadly, and our social safety net, as some individuals see their jobs becoming automated. Even, and especially those, who don't ever see their jobs as the type that could become automated.

If you ask most people if they are likely to be replaced by robots, most people are probably going to say no. A much higher percentage of them are actually likely to see their jobs replaced, perhaps within their lifetimes.

The American economy has been utterly transformed in the decades since we first created most of what we think of as our first safety net. That is not just welfare programs, but also our education system, and our health care system. In that time, Washington has not kept up. And now as these disruptions are accelerating, policy is falling much further behind and the American people are paying a price for Washington's systemic and chronic policy paralysis.

The nature of work is changing, but the nature of human beings is not. Work, vocation, service, providing, is at the nature and is at the core of human happiness. It is a source of meaning and dignity and community, and that goes for all of us, not just for people with college degrees, and with skills that put us into an elite category.

The American people have been resilient, and they have been more than patient, but they are competing in a 21st century economy while relying on a fraying 20th century safety net. None of us has all the answers. And I think that is the point. We need a safety net policy, or a set of policies, that are as flexible and nimble and diverse and as adaptable to technological change as our society and economy are becoming. And this is something that is going to need the insights of both parties to get it right—and not just both parties, but the insights of the brightest minds in our country. And for that reason, we have been very grateful to have you here to provide that.

As we have discussed today, we cannot ignore the broader impact automation will have on our society and the policy changes that Congress should be considering as the nature of work continues to change, as it has been for centuries, but as it appears to be doing on an accelerated basis.

Again, thank you for coming. Thank you for your participation and your insight today. And in closing out the hearing I would be remiss if I didn't remind everyone that the gavel I am using to close out the hearing, this blue gavel, was made by the good students at Washington Math and Science Tech Public High School.

Thank you, very much. We will stand adjourned.

(Whereupon, at 4:25 o'clock p.m., Wednesday, May 25, 2016, the hearing was adjourned.)

SUBMISSIONS FOR THE RECORD

PREPARED STATEMENT OF HON. DANIEL COATS, CHAIRMAN, JOINT ECONOMIC COMMITTEE

Today the Committee will examine how robots, automation, and technology are transforming our economy. I'd like to thank our witnesses for being here, and I will be introducing them shortly.

But first, I would like to draw the Committee's attention to the gavel I just used to start today's hearing. It looks and functions just like a typical gavel, but it was not made from a block of wood. In fact, it was created just down the street at the Washington, D.C., Public Library's Fabrication Laboratory, or "Fab Lab," using 3D printing.

3D printing works by heating up raw material, in this case plastic, and "printing" one small layer at a time until the object is completed. Rather than needing to mold or carve raw material as in the past, now we can simply upload a file to a printer and it will create the item according to the user's exact specifications.

I also have with me a different 3D-printed gavel that we will use to adjourn this hearing. It was made by students at the Washington Mathematics Science Technology Public Charter High School, also located in the District of Columbia. What an exciting new world we live in, where objects can be manufactured on demand with such ease and specificity.

I would like to thank both institutions for their contributions to today's hearing, which tangibly illustrate the topic we are about to explore. I would also like to thank Senator Lee and his staff for helping the Committee prepare for today's hearing.

Recent technological developments have been pushing the envelope faster and further than was expected even a decade ago, making what was once thought of as science fiction a reality. I remember the hassle of getting my children to program our VCR. Now my cable box is capable of recording all my favorite shows, without me even asking. Meanwhile some of my grandchildren are probably asking, "What is a VCR?"

The robotic machines are here. Whether it is vacuuming our carpets or assisting in precise surgeries, robots are helping with and performing almost any task we can imagine. This has led to a greater abundance of consumer products, and more productive and creative workers.

However, as with the Industrial Revolution, this new robot revolution clearly is contributing to pressures arising within our changing labor force. Even before these technological advances, America's workforce was starting to age and businesses were beginning to rely much more on automated labor than physical labor. Robots are expected to hasten this trend as they fill in for humans in both blue- and white-collar jobs.

This picture of a modern assembly line illustrates the prevalence of automation in today's economy. Where workers used to assemble vehicles directly by hand, now they oversee teams of precise robots that can weld and assemble vehicles far more advanced than ever before.

Automation's rapid progress has also raised challenges with certain government policies. How can we foster an environment where innovators thrive and grow? Is our social safety net prepared for a 21st century labor market? Do some government policies make human workers prohibitively expensive for employers? How will current workers adapt? And is our education system preparing our youngest citizens for the future economy?

These are important questions. For guidance, we look forward to hearing the views of our distinguished witnesses.

Today we will hear from Dr. Andrew McAfee, principal research scientist and co-founder of MIT's Institute Initiative on the Digital Economy. We also welcome Adam Keiper, fellow at the Ethics and Public Policy Center and editor of the quarterly technology publication The New Atlantis. Our final witness is Harry Holzer, professor at the McCourt School of Public Policy at Georgetown University and a Senior Fellow in Economic Studies at the Brookings Institution.

My thanks to all of you for providing us with your expertise and giving us a glimpse into the possibilities of the future.

PREPARED STATEMENT OF CAROLYN B. MALONEY, RANKING DEMOCRAT, JOINT ECONOMIC COMMITTEE

Thank you so much Chairman Coats for calling such an important and interesting and timely hearing. We are here today to discuss the impact of automation on jobs and the economy and how best to harness the immense power of technological innovation.

The United States has long been a leader in this important area. And basic research funded by the federal government has played a key role in driving innovation.

We know that automation can boost productivity, lift aggregate demand, reduce consumer prices and improve our quality of life.

While all of these benefits are apparent in the long run, we also know that in the short run innovation can displace workers, causing severe economic pain to workers whose jobs are automated out of existence or whose wages are reduced dramatically.

Today's hearing is about the future. And let's face it: automation is a difficult thing to predict. We know it's going to happen. We just don't know how fast it's going to happen, or in which industries or what will be the exact consequences.

One study finds that nearly half of U.S. jobs are at risk of being lost to automation in the next couple of decades.

Other studies show that the impacts of automation will not be as great or felt so soon.

Throughout history, concerns have been voiced that new technologies would make human labor obsolete. It has not happened. While there have been dramatic shifts in how people have earned their livings, the quantity of jobs has increased and the quality has improved.

Yet, there are reasons to believe that this could be different in the future.

I'd like to add some of my questions to the excellent questions Senator Coats put forward.

- How do we equip our workers with the tools and skills needed to adapt to the future changes?
- What should we do as policymakers to both advance innovation and the expected productivity benefits on the one hand while also supporting workers adversely affected by technological changes on the other hand?
- And how can we harness this engine of prosperity while making sure that benefits are widely shared?

I really am excited to learn more and to hear the questions and exchange here today with our excellent witnesses.

But before I yield back my time, I'd like to turn to Senator Peters, a former colleague in the House of Representatives, we miss you. And I'd like to yield the balance of my time to him. He is the co-founder of the bipartisan Senate Smart Transportation Caucus. Senator Peters has a deep interest and knowledge of automation and its impacts in Michigan and the rest of the United States.

I yield him the remainder of my time, and it's always good to see you again.

Four Points about Robots, Jobs, and Wages

Testimony of Andrew McAfee before the Joint Economic Committee
hearing on "The Transformative Impact of Robots and Automation"
May 25, 2016

Andrew McAfee, MIT
amcafee@mit.edu, @amcafee

Four Points about Robots, Jobs, and Wages

I would like to thank Chairman Coats and Ranking Member Maloney for giving me the opportunity to testify today on this important topic.

I'd like to make four points. First, that the American workforce is undergoing substantial and important changes. Second, that technological progress is one of the driving forces behind these changes. Third, that tech progress is accelerating at present, and that this speedup will have consequences for the labor force. And fourth, that effective policy responses to these changes do exist and should be put in place.

Important recent changes in the American economy can be seen in the graph below, which illustrates a phenomenon my coauthor Erik Brynjolfsson and I call "the great decoupling." It is a decoupling between measures of output, like GDP per capita and labor productivity, and measures of the health of the workforce, such as job volume and median household income. Output measures have risen fairly steadily throughout the post-war era, declining only during recessions (of which the most recent is also the most significant). Workforce measures used to rise almost in lockstep with output measures, but as the graph shows in recent decades the pattern has changed, with job and income growth both tapering off. By some measures, in fact, median American household income is lower than it was at the turn of the century.

Four Points about Robots, Jobs, and Wages

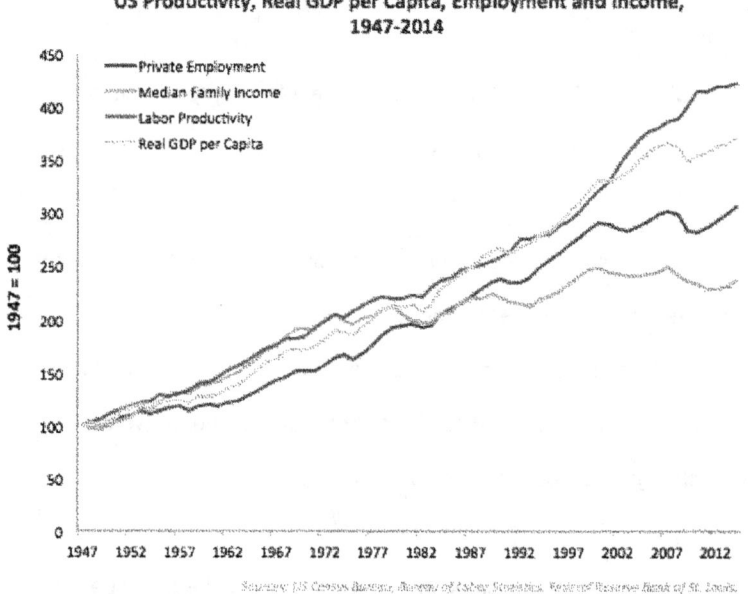

Other research, much of it conducted by my MIT colleague David Autor and his co-authors, shows that the historically large, stable, and prosperous American middle class has in recent decades been getting "hollowed out." High-wage jobs, like corporate executive or computer programmer, are still being created, as are low-wage ones like restaurant busboy and home health aide. However, mid-wage, mid-skill jobs are not being created at the same rate as in the past. Many of these jobs involve routine work, both physical and cognitive; the archetypal routine physical job was on a factory assembly line, while a typical routine knowledge work job was as a payroll clerk working at that same factory. Jobs like these are becoming increasingly rare within the American workforce.

Four Points about Robots, Jobs, and Wages

Why is this? There is an active debate at present about the reasons for the great decoupling and the hollowing out of the workforce. My read of the available evidence is that two causes are particularly important: globalization and technological progress. The increasing globalization of trade in recent decades and the opening up of large economies like China and India have meant that both physical work and knowledge work that used to be done in America have moved overseas. At the same time, technological progress has allowed more physical and knowledge work to be automated.

Both trade and technology have had their greatest impact on routine work. Many assembly lines have moved overseas, and many of the ones that have remained in the United States have become highly automated. Much routine "back office" knowledge work like processing payments also now takes place overseas in low-wage countries, or entirely inside a computer.

It is important to stress that both global trade and technological progress have been hugely beneficial to Americans and the American economy as a whole. Both have increased our overall prosperity, given us access to previously unavailable goods and services at lower prices and higher levels of quality, and benefited many American workers by opening up new markets and furnishing better tools to help them in their jobs.

However, there is no economic law that says that the benefits from either trade or tech progress must be shared equally, or in a way that strikes us as "fair." It is perfectly possible for many of these benefits to go to a relatively small group. When economists speak about the "distributional consequences" of trade or technological progress, this is what they mean. The consequence that most concerns me is that as both trade and

Four Points about Robots, Jobs, and Wages

technology race ahead, it seems that they are leaving many Americans behind in their capacity as people who want to offer their labor to an employer in exchange for a good standard of living.

The impact of technological progress on employment is clearly visible in the history of America's manufacturing Industries. The graph below shows total domestic manufacturing output along with total us manufacturing employment. It reveals that overall manufacturing employment has been declining since about 1980, even as the sector's output has continued to rise. This is simply a story of increased productivity, which in turn is largely a story of automation and technological progress.

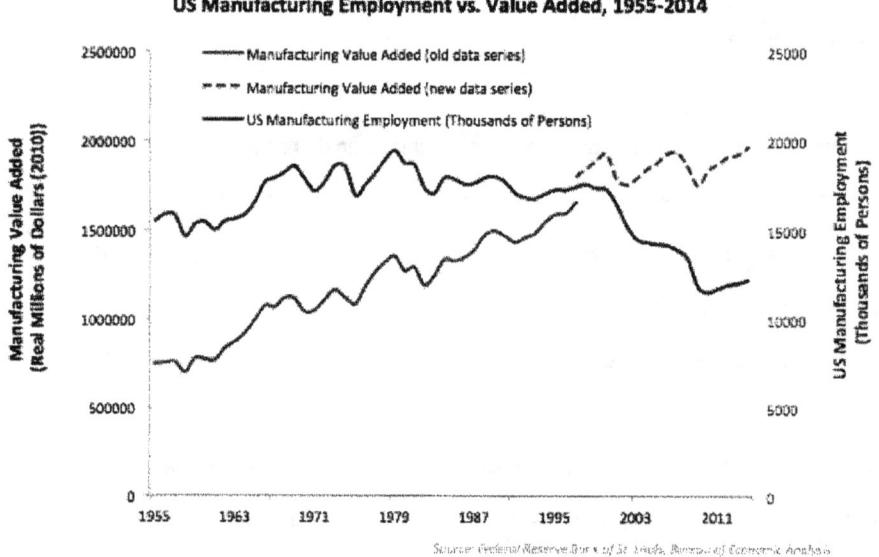

My third point Is that we are living in a time when this progress is accelerating rapidly. The promises of science fiction are becoming everyday economic reality. We have cars

Four Points about Robots, Jobs, and Wages

and trucks that can drive themselves; swarms of autonomous drones that can work together to survey a construction site or build a rope bridge; software that can understand our spoken questions and give us accurate answers; machines that can print three-dimensional parts out of plastic, ceramic, or metal; thermostats that learn over time when we're home and adjust the heating and air-conditioning accordingly; and artificial intelligence systems that can quite literally beat us at our own games: poker, chess and the ancient Asian strategy game of go.

One particularly important recent advance, which is often described under the label "machine learning," is our ability to build systems that don't need to be told by their human programmers exactly what steps to follow in order to achieve a desired result. Instead, these systems learn by being shown lots of examples (both of successes and failures) and eventually figure out the relevant rules, strategies, and patterns by themselves. This means that for the first time we humans have a digital colleague or second opinion that can, like us, look at information and draw inferences and conclusions from it. These systems can take in much more information than even an expert human can, which opens up intriguing possibilities. I am convinced, for example, that if the world's best medical diagnostician is not today a piece of technology, it soon will be.

Many of these innovations are quite new, and have not yet diffused throughout the economy. As they do, they will intensify technology's already substantial effects on the workforce. In other words, they will benefit some types of workers, and substitute for others. I do not believe that the era of mass technological unemployment is right around the corner, or, in other words, that the robots are about to take all of our jobs, But I

Four Points about Robots, Jobs, and Wages

believe the technology surge now underway will intensify the great decoupling, the hollowing out of the middle class, and other trends already underway.

My last point concerns the steps that the government can and should take to continue to reap the benefits of technological progress while dealing with the challenges it poses for some workers. I believe this is possible, yet we're not doing it.

For me and many others, the most frustrating part of the current economic policy environment is that we're not getting right the "Economics 101 playbook" — the set of things that virtually all decent economists, conservative and liberal, agree on. My playbook for the near future has five main elements, and to remember them I sing to myself the old nursery rhyme about Old McDonald's farm. The origin of the song's "E-I-E-I-O" refrain is unclear, but for me it stands for education, infrastructure, entrepreneurship, immigration and original research.

Education: The primary school system in the US has been called the country's best idea, but at present the country's students are no better than middle of the pack internationally. There is alarming evidence that college students are often learning very little and there's still too much focus on rote learning and mastering skills that technology is already quite good at.

Infrastructure: World-class roads, airports and networks are investments in the future and a cornerstone of strong growth. But the American Society of Civil Engineers gives the US an overall grade of D+, and internet speeds here are slower than in many other countries.

Four Points about Robots, Jobs, and Wages

Entrepreneurship: Young businesses, especially fast-growing ones, are a prime source of new jobs. Unfortunately, entrepreneurship and business dynamism in the US have been on a slow, steady decline. No one knows exactly why this is - the economist John Haltiwanger has characterized the situation as "death by a thousand cuts" - but many believe that the thicket of regulations, licensing requirements, and other barriers confronting job creators is a serious impediment. American innovators and entrepreneurs are the best hope for American workers, and we need to make life easier instead of increasingly difficult for them.

Immigration: Many of the world's most talented and ambitious people want to come to the US to build their lives and careers, and the evidence is clear that immigrant-founded companies have been great job-creation engines. Yet our policies in this area are far too restrictive, and our procedures are nightmarishly bureaucratic.

Original research: Companies tend to concentrate on applied research where they can capture the rewards from their efforts. This means that government has a role to play in supporting original, early-stage work for which the rewards are spread more broadly. Most of today's tech marvels, from the internet to the smartphone, have a government programme somewhere back in their family tree. But funding for basic research in the US is on the decline as a share of gross domestic product.

We live in interesting times. I believe that modern digital technologies are poised to reshape our economy as profoundly as the combination of electricity and the internal combustion engine did a century ago. These new tools will greatly increase the productivity and abundance of our economy, and the overall prosperity of our people. It is up to us to shape how these benefits are shared. I'll close by repeating the last

Four Points about Robots, Jobs, and Wages

sentence of our book *The Second Machine* Age: technology is not destiny; we shape our destiny.

Testimony Presented to the Joint Economic Committee:

The Transformative Impact of Robots and Automation

Adam Keiper
Fellow, Ethics and Public Policy Center
Editor, *The New Atlantis*
May 25, 2016

Mr. Chairman, Ranking Member Maloney, and members of the committee, thank you for the opportunity to participate in this important hearing on robotics and automation. These aspects of technology have already had widespread economic consequences, and in the years ahead they are likely to profoundly reshape our economic and social lives.

Today's hearing is not the first time Congress has discussed these subjects. In fact, in October 1955, a subcommittee of this very committee held a hearing on automation and technological change.[1] That hearing went on for two weeks, with witnesses mostly drawn from industry and labor. It is remarkable how much of the public discussion about automation today echoes the ideas debated in that hearing. Despite vast changes in technology, in the economy, and in society over the past six decades, many of the worries, the hopes, and the proposed solutions suggested in our present-day literature on automation, robotics, and employment would sound familiar to the members and witnesses present at that 1955 hearing.

It would be difficult to point to any specific policy outcomes from that old hearing, but it is nonetheless an admirable example of responsible legislators grappling with immensely complicated questions. A free people must strive to govern its technologies and not passively be governed by them. So it is an honor to be a part of that tradition with today's hearing.

---

[1] "Automation and Technological Change," hearings before the Subcommittee on Economic Stabilization of the Joint Committee on the Economic Report, Congress of the United States, Eighty-fourth Congress, first session, October 14, 15, 17, 18, 24, 25, 26, 27, and 28, 1955 (Washington, D.C.: G.P.O., 1955), http://www.jec.senate.gov/public/index.cfm/1956/12/report-970887a6-35a4-47e3-9bb0-c3cdf82ec429.

In my remarks, I wish to make five big, broad points, some of them obvious, some more counterintuitive.

(1) WHY IT IS SO HARD TO KNOW THE FUTURE

A good place to start discussions of this sort is with an admission of humility. When reviewing the mountains of books and magazine articles that have sought to predict what the future holds in automation and related fields, when reading the hyped tech headlines or when looking at the many charts and tables extrapolating from the past to help us forecast the future, it is striking to see how often our predictions go wrong.

Very little energy has been invested in systematically understanding why futurism fails — that is, why, beyond the simple fact that the future hasn't happened yet, we have generally not been very good at predicting what it will look like. For the sake of today's discussion, I want to raise just a few points, each of which can be helpful in clarifying our thinking when it comes to automation and robotics.

First there is the problem of timeframes. Very often, economic analyses and tech predictions about automation discuss kinds of jobs that are likely to be automated without any real discussion of *when*. This leads to strange conversations, as when one person is interested in what the advent of driverless vehicles might mean for the trucking industry, and his interlocutor is more interested in, say, the possible rise of artificial superintelligences that could wipe out all life on Earth. The timeframes under discussion at any given moment ought to be explicitly stated.

Second there is the problem of context. Debates about the future of one kind of technology rarely take into account other technologies that might be developed, and how those other technologies might affect the one under discussion. When one area of technology advances, others do not just stand still. How might automation and robotics be affected by developments in energy use and storage, or advanced nanotechnology (sometimes also called molecular manufacturing), or virtual reality and augmented reality, or brain-machine interfaces, or various biotechnologies, or a dozen other fields?

And of course it's not only other technologies that evolve. In order to be invented, built, used, and sustained, all technologies are enmeshed in a web of cultural practices and mores, and legal and political norms. These things do not stand still either — and yet when discussing the future of a given technology, rarely is attention paid to the way these things touch upon one another.

All of which is to say that, as you listen to our conversation here today, or as you read books and articles about the future of automation and robotics, try to avoid making the leaps of logic that so often throw off futurist speculation — what I call the "chain of uncertainties":

Just because something is conceivable or imaginable

does not mean it is possible.

Even if it is possible, that does not mean it will happen.

Even if it happens, that does not mean it will happen in the way you envisioned.

And even if it happens in something like the way you envisioned, there will be unintended, unexpected consequences

(2) WHY THIS TIME IS DIFFERENT

Automation is not new. For thousands of years we have made tools to help us accomplish difficult or dangerous or dirty or tedious or tiresome tasks, and in some sense today's tools are just extensions of what came before. And *worries* about automation are not new either — they date back at least to the early days of the Industrial Revolution, when the Luddites revolted in England over the mechanization and centralization of textile production. As I mentioned above, this committee was already discussing automation some six decades ago — thinking about thinking machines and about new mechanical modes of manufacturing.

What makes today any different?

There are two reasons today's concerns about automation are fundamentally different from what came before. First, the *kinds of "thinking"* that our machines are capable of doing is changing, so that it is becoming possible to hand off to our machines ever more of our cognitive work. As computers advance and as breakthroughs in artificial intelligence (AI) chip away at the list of uniquely human capacities, it becomes possible to do old things in new ways and to do new things we have never before imagined.

Second, we are also *instantiating intelligence* in new ways, creating new kinds of machines that can navigate and move about in and manipulate the physical world. Although we have for almost a century imagined how robotics might transform our world, the recent blizzard of technical breakthroughs in movement, sensing, control, and power is bringing us for the first time into a world of autonomous, mobile entities that are neither human nor animal.

To simplify a vast technical and economic literature, there are basically three futurist scenarios for what the next several decades hold in automation, robotics, and artificial intelligence:

Scenario 1 – Automation and artificial intelligence will continue to advance, but at a pace sufficiently slow that society and the economy can gradually absorb the changes, so that people can take advantage of the new possibilities without suffering the most disruptive effects. The job market will change, but in something like the way it has changed over the last half-century: some kinds of jobs will disappear, but new kinds of jobs will be created, and by and large people will be able to adapt to the shifting demands on them while enjoying the great benefits that automation makes possible.

Scenario 2 – Automation, robotics, and artificial intelligence will advance very rapidly. Jobs will disappear at a pace that will make it difficult for the workforce to adapt without widespread difficulty. The kinds of jobs that will be threatened will increasingly be jobs that had been relatively immune to automation — the "high-skilled" jobs that generally involved creativity and problem-solving, and the "low-skilled" jobs that involved manual dexterity or some degree of adaptability and interpersonal relations. The pressures on low-skilled American workers will exacerbate those already felt because of competition against foreign workers paid lower wages. Among the disappearing jobs may be those at the lower-wage end of the spectrum that we have counted on for decades to instill basic workplace skills and values in our young people, and that have served as a kind of employment safety net for older people transitioning in their lives. And the balance between labor and capital may (at least for a time) shift sharply in favor of capital, as the share of gross domestic product (GDP) that flows to the owners of physical capital (e.g., the owners of artificial intelligences and robots) rises and the share of GDP that goes to workers falls. If this scenario unfolds quickly, it could involve severe economic disruption, perhaps social unrest, and maybe calls for political reform. The disconnect between productivity and employment and income in this scenario also highlights the growing inadequacy of GDP as our chief economic statistic: it can still be a useful indicator in international competition, but as an indicator of economic wellbeing, or as a proxy for the material satisfaction or happiness of the American citizen, it is clearly not succeeding.

Scenario 3 – Advances in automation, robotics, and artificial intelligence will produce something utterly new. Even within this scenario, the range of possibilities is vast. Perhaps we will see the creation of "emulations," minds that have been "uploaded" into computers. Perhaps we will see the rise powerful artificial "superintelligences," unpredictable and dangerous. Perhaps we will reach a "Singularity" moment after which everything that matters most will be different from what came before. These types of possibilities are increasingly matters of discussion for technologists, but their very radicalness makes it difficult to say much about what they might mean at a human scale — except insofar as they might involve the extinction of humanity as we know it.

One can make a plausible case for each of these three scenarios. But rather than discussing their likelihood or examining some of the assumptions and aspirations inherent in each scenario, in the limited time remaining, I am going to turn to three other broad subjects: some of the legal questions raised by advances in artificial intelligence and automation; some of the policy ideas that have been proposed to mitigate some of the anticipated effects of these changes; and a deeper understanding of the meaning of work in human life.

(3) LOOMING LEGAL QUESTIONS

The advancement of artificial intelligence and autonomous robots will raise questions of law and governance that scholars are just beginning to grapple with. These questions are likely to have growing economic and perhaps political consequences in the years to come, no matter which of the three scenarios above you consider likeliest.

The questions we might be expected to face will emerge in matters of liability and malpractice and torts, property and contractual law, international law, and perhaps laws related to legal personhood. Although there are precedents — sometimes in unusual corners of the law — for some of the questions we will face, others will arise from the very novelty of the artificial autonomous actors in our midst.

By way of example, here are a few questions, starting with one that has already made its way into the mainstream press:

- When a self-driving vehicle crashes into property or harms a person, who is liable? Who will pay damages?

- When a patient is harmed or dies during a surgical operation conducted by an autonomous robotic device upon the recommendation of a human physician, who is liable and who pays?

- If a robot is autonomous but is not considered a person, who owns the creative works it produces?

- In a combat setting, who is to be held responsible, and in what way, if an autonomous robot deployed by the U.S. military kills civilian noncombatants in violation of the laws of war?

- Is there any threshold of demonstrable achievement — any performed ability or set of capacities — that a robot or artificial intelligence could cross in order to be entitled to legal personhood?

These kinds of questions raise matters of justice, of course, but they have economic implications as well — not only in terms of the money involved in litigating cases, but in terms of the effects that the legal regime in place will have on the further development and implementation of artificial intelligence and robotics. It will be up to lawyers and judges, and lawmakers at the federal, state, and local levels, to work through these and many other such matters.

(4) PROPOSED SOLUTIONS AND THEIR PROBLEMS

There are, broadly speaking, two kinds of ideas that have most often been set forth in recent years to address the employment problems that may be created by an increasingly automated and AI-dominated economy.

The first category involves adapting workers to the new economy. The workers of today, and even more the workers of tomorrow, will need to be able to pick up and move to where the jobs are. They should engage in "lifelong learning" and "upskilling" whenever possible to make themselves as attractive as possible to future employers. Flexibility must be their byword.

To be sure, in a churning free economy, some sort of flexibility is always a good thing; flexibility can mean resilience in times of creative destruction. Yet we must remember that "workers" are not just workers; they are not just individuals free and detached and able to go wherever and do whatever the market demands. They are also members of families — children and parents and siblings and so on — and members of communities, with the web of connections and ties those memberships imply. And maximizing flexibility can be detrimental to those kinds of relationships, relationships that are necessary for human flourishing.

The other category involves a universal basic income — or what is sometimes called a "negative income tax" — guaranteed to every individual, even if he or she does not work. This can sound, in our contemporary political context, like a proposal for redistributing wealth, and it is true that there are progressive theorists and anti-capitalist activists who support it. But this idea has also been discussed favorably for various reasons by prominent conservative and libertarian thinkers. It is an intriguing idea, and one without many real-life models that we can study (although Finland is currently contemplating an interesting partial experiment).

A guaranteed income certainly would represent a sea change in our nation's economic system and a fundamental transformation in the relationship between citizens and the state, but perhaps this transformation would be suited to the technological challenge we may face in the years ahead. Some of the smartest and most thoughtful analysts have discussed how to avoid the most obvious problems a guaranteed income might create — such as the problem of disincentivizing work. Especially provocative is the depiction of guaranteed income that appears in a 2008 book written by Joseph V. Kennedy, a former senior economist with the Joint Economic Committee; in his version of the policy, the guaranteed income would be structured in such a way as to encourage a number of good behaviors. Anyone interested in seriously considering guaranteed income should read Kennedy's book.[2]

(5) THE MEANING OF HUMAN WORK

Should we really be worrying so much about the effects of robots on employment? Maybe with the proper policies in place we can get through a painful transition and reach a future date when we no longer *need* to work. After all, shouldn't we agree with Arthur C. Clarke that "The goal of the future is full *un*employment"?[3] Why work?

This notion, it seems to me, raises deep questions about who and what we are as human beings, and the ways in which we find purpose in our lives. A full discussion of this subject would require drinking deeply of the best literary and historical investigations of work in human life — examining how work is not only a

---

[2] Joseph V. Kennedy, *Ending Poverty: Changing Behavior, Guaranteeing Income, and Reforming Government* (Lanham, Md.: Rowman and Littlefield, 2008).

[3] Arthur C. Clarke, quoted by Jerome Agel, "Cocktail Party" (column), *The Realist* 86, Nov.–Dec. 1969, page 32, http://ep.tc/realist/86/32.html.

matter of toil for which we are compensated, but how it also can be a source of dignity, structure, meaning, friendship, and fulfillment.

For present purposes, however, I want to just point to two competing visions of the future as we think about work. Because, although science fiction offers us many visions of the future in which man is *destroyed* by robots, or *merges* with them to become cyborgs, it offers basically just two visions of the future in which man *coexists* with highly intelligent machines. Each of these visions has an implicit anthropology — an understanding of what it means to be a human being. In each vision, we can see a kind of liberation of human nature, an account of what mankind would be in the absence of privation. And in each vision, some latent human urges and longings emerge to dominate over others, pointing to two opposing inclinations we see in ourselves.

The first vision is that of the techno-optimist or -utopian: Thanks to the labor and intelligence of our machines, all our material wants are met and we are able to lead lives of religious fulfillment, practice our hobbies, pursue our intellectual and creative interests.

Recall John Adams's famous 1780 letter to Abigail: "I must study Politicks and War that my sons may have liberty to study Mathematicks and Philosophy. My sons ought to study Mathematicks and Philosophy, Geography, natural History, Naval Architecture, navigation, Commerce and Agriculture, in order to give their Children a right to study Painting, Poetry, Musick, Architecture, Statuary, Tapestry and Porcelaine."[4] This is somewhat like the dream imagined in countless stories and films, in which our robots make possible a Golden Age that allows us to transcend crass material concerns and all become gardeners, artists, dreamers, thinkers, lovers.

By contrast, the other vision is the one depicted in the 2008 film *WALL-E*, and more darkly in many earlier stories — a future in which humanity becomes a race of Homer Simpsons, a leisure society of

---

[4] John Adams to Abigail Adams (letter), May 12, 1780, Founders Online, National Archives (http://founders.archives.gov/documents/Adams/04-03-02-0258). Source: *The Adams Papers*, Adams Family Correspondence, vol. 3, *April 1778–September 1780*, eds. L. H. Butterfield and Marc Friedlaender (Cambridge, Mass.: Harvard, 1973), pages 341–343.

consumption and entertainment turned to endomorphic excess. The culminating achievement of human ingenuity, robotic beings that are smarter, stronger, and better than ourselves, transforms us into beings dumber, weaker, and worse than ourselves. TV-watching, video-game-playing blobs, we lose even the energy and attention required for proper hedonism: human relations wither and natural procreation declines or ceases. Freed from the struggle for basic needs, we lose a genuine impulse to strive; bereft of any civic, political, intellectual, romantic, or spiritual ambition, when we do have the energy to get up, we are disengaged from our fellow man, inclined toward selfishness, impatience, and lack of sympathy. Those few who realize our plight suffer from crushing ennui. Life becomes nasty, brutish, and long.

Personally, I don't think either vision is quite right. I think each vision — the one in which we become more godlike, the other of which we become more like beasts — is a kind of deformation. There is good reason to challenge some of the technical claims and some of the aspirations of the AI cheerleaders, and there is good reason to believe that we are in important respects stuck with human nature, that we are simultaneously beings of base want and transcendent aspiration; finite but able to conceive of the infinite; destined, paradoxically, to be free.

CONCLUSION

Mr. Chairman, the rise of automation, robotics, and artificial intelligence raises many questions that extend far beyond the matters of economics and employment that we've discussed today — including practical, social, moral, and perhaps even existential questions. In the years ahead, legislators and regulators will be called upon to address these technological changes, to respond to some things that have already begun to take shape and to foreclose other possibilities. Knowing when and how to act will, as always, require prudence.

My hope is that, as we contemplate the blessings and the burdens of these new technologies, we will resist the temptation to relinquish responsibility to our creations, and we will strive always to point their use in a humane direction.

Testimony of Harry J. Holzer before the Joint Economic Committee of the US Congress

May 25, 2016

The Effects of Automation on US Labor Markets and Policy

I'd like to make a number of points about how technology and automation will affect the US labor market, and the implications of those effects for a range of labor market policies.

1. *Fears of how automation eliminates jobs have historically been greatly overblown.*

As far back as the Luddites in Britain, and at other times in the US, workers have feared that technology would eliminate millions of jobs and cause mass unemployment. This has never turned out to be true. Markets have ways of adjusting to technology that create new jobs – specifically, as worker productivity rises and prices decline, consumers' real incomes rise, and they spend more on other goods and services, creating new jobs in these sectors. Indeed, a century of dramatic productivity growth from the late 19th through the late 20th century in the US generated no aggregate job loss in the long run.[i] But workers in the specific jobs and sectors directly affected by technology often are displaced from those jobs, and experience lengthy periods of unemployment and reduced wages when they ultimately become employed.

2. *While technology hasn't eliminated large numbers of jobs in the aggregate, it can reduce earnings among large groups of workers.*

Even among workers who are not directly displaced by technology in the workplace, labor market demand for their skills can be reduced. In the past 35 years, the digital revolution – among other factors, such as globalization and weakening institutions like unions – has reduced employment in many good-paying job categories (or reduced wages in those that remain) for workers with high school or less education. The jobs affected have been mostly in goods production among men and clerical work among women, since these involve routine tasks that are most easily replaceable by the new technologies; unfortunately, the new jobs available to them in the service sector pay considerably less than those eliminated.

At the same time, wages and jobs increase for workers with the technical skills to use the new technology (such as engineers, machinists and other technicians) or other skills that complement the new machines - including analytical, communications or creative skills. In other words, technical change has a "skill bias" in the labor market, with relatively unskilled workers hurt by it while more skilled workers are helped. In addition, there seems to be a "capital bias" as well, with the owners of businesses that use the new technologies enhancing their share of national income at the expense of workers more broadly.

Within the labor market, the skill bias in technical change causes growing "polarization" of jobs between the low-paying and high-paying sectors. The middle of the job market is not really disappearing; but newer middle-paying jobs - like those in health care, IT, advanced manufacturing and many parts of the service sector – require more postsecondary education or training (though short of a BA) than did the earlier production and clerical jobs.

And the growth in this "newer middle" is not sufficiently large to offset the decline in the "older middle" of production and clerical jobs, leading to some "hollowing out" of the middle of the labor market overall. Specifically, between 2000 and 2013, the share of all jobs accounted for by the "older middle" shrank from 24.3 to 21 percent while those of the "newer middle" grew from 14.8 to 15.6 percent. Thus, the shares of all jobs in the middle-paying category shrank from 39.1 to 36.6 percent.[ii]

The polarization in the job market has contributed to stagnating or declining real wages for unskilled workers, plus dramatic increases in earnings inequality. For instance, real earnings for American men with high school diplomas or less declined by over 10 percent between 1979 and 2012, while those for workers with BA and graduate degrees increased by about 20 and 70 percent respectively. Stagnant or declining wages of less-educated men, in turn, reduce their labor force participation as well as their marriage rates, thereby hurting not only the overall economy but also families and communities.[iii]

3. *The future effects of "artificial intelligence" and robotics in the workplace are very hard to predict, though the breadth and pace of labor market dislocations could grow.*

It will be a long time before we know the labor market effects of the next generation of robots and other digital technologies in the workplace, as it often takes decades for employers to figure out how to use them efficiently. At least in theory, the threats of job displacement could widen over time, and threaten millions more workers than it has so far; and large-scale displacement could potentially overwhelm the market adjustment mechanisms described above, creating years of sluggish demand in the labor market. At least to date, we have seen little evidence of this, outside a few key sectors (like manufacturing); if anything, US labor markets have become *less* fluid and dynamic over the past few decades, and our productivity growth in the past decade has sagged.[iv] But, over the next few decades, the pace and breadth of dislocations could grow, as new technologies are generated and employers gradually learn how to use them in the workplace more effectively. Though productivity and therefore worker incomes will grow as a result, jobs could become more unstable and harder to find among workers of all skill levels than before.

4. *Future automation should NOT become an excuse to avoid or eliminate a sensible and moderate set of worker supports and services that help them address the labor market challenges described above.*

The skill bias of new technologies means that workers will need to gain new skills to improve their wages and reduce inequality, while we also try to "make work pay" somewhat more for unskilled workers to keep them in the labor force. The capital bias might also imply a need to raise or supplement wages more broadly. Rising displacements and job instability create a need for important benefits like health care and family/medical leave to be portable across jobs and available during period of unemployment. And, if displacements outpace the new job creation rate in the future, we might need policies to spur labor demand and create more jobs.

Fear that providing these job market supports might raise costs to employers, and therefore lead to faster mechanization over time, have little merit as long as the supports in question are *moderate* in magnitude, and especially if they are offset by workers whose skills and productivity are enhanced.

The needed range of policies and supports to deal with the potential costs of automation include the following:

A. Raising/Protecting Worker Earnings from Skill and Capital Biases of Technology
- *Education and Skill Development* - Clearly, support for and reforms in public programs and institutions (like community colleges) are needed to improve the skills of US workers, and help them adapt over time to changes in skill demands in a dynamic labor market. We need more workforce services like career counseling and job search assistance, community college training that is more responsive to the labor market, newer models of high-quality career and technical education plus work-based learning (e.g., apprenticeships), and opportunities for life-long learning that would enable displaced workers to upgrade or change their skill sets over time.
- *Protecting Worker Rights to Collective Bargaining* – The current legal assault on unions in both the public and private sectors will weaken collective bargaining and further exacerbate wage inequality and earnings stagnation, with its resulting declines in labor force activity and family formation.
- *Supporting High-Road Job Creation by Private Employers* - Governments at all levels could commit to creating "good jobs" and "high-performance workplaces" by rewarding and assisting employers who invest in skill-upgrading and improving the productivity and compensation of their workers through apprenticeships, incumbent worker training, profit-sharing, and other such mechanisms.[v] This would improve economic productivity in the US while providing important gains to workers and their families, without reducing profits for companies.

B. Making Work Pay for the Unskilled
- *Wage Insurance against Displacement and the Earned Income Tax Credit (EITC)* - Expansions in wage insurance for displaced workers, and in the EITC for low-income workers in general, would incentivize them to accept newer jobs that pay less. These policies would likely raise labor force participation among those who have been dropping out in recent years.
- *Minimum Wage Increases* - Increases in the minimum wage would also help to "make work pay" and would reduce reliance on other income supports like food stamps and Medicaid. As long as they are *moderate* in magnitude and introduced gradually, they should not accelerate the potential mechanization of jobs in "fast food" or other retail sectors.

C. Protecting Workers from a More Unstable Job Market
- *Portable Health/Family Benefits* - The Affordable Care Act helps many millions of low-skill workers obtain health insurance while likely reining in per-capita increases in health care costs. Health benefits are also becoming more portable, so workers retain them even when they lose jobs. In addition, paid family and medical leave – funded through payroll taxes rather than mandates on employers – would help parents of small children (particularly mothers) and care-giving adults remain attached to

their jobs and the workforce while they deal with important personal or family needs. Investments of parental time in their children raises worker productivity over time.

D. Creating More Jobs
- *Public Funds for Public or Private Job Creation* – If/when new technologies lead to large worker dislocations that outpace the labor market's ability to create new jobs, we might need to supplement job creation. For instance, sensible public spending on infrastructure would help fix our crumbling roads and bridges, thus increasing economic productivity growth, while bolstering labor demand when needed. Subsidizing jobs more broadly in the public and private sectors, which we did during the Great Recession, can successfully spur net employment among disadvantaged workers and help meet employer needs as well.[vi]

---

[i] Robert J. Gordon, *The Rise and Fall of American Growth: The US Standard of Living Since the Civil War*. Princeton University Press, 2015.

[ii] Harry J. Holzer, *Job Market Polarization and Worker Skills: A Tale of Two Middles*. Brookings Institution, 2015.

[iii] David Autor, "Skills, Education, and the Rise of Earnings Inequality among the "other 99 percent." *Science*, 2014; and David Autor et al. The Labor Market and the Marriage Market: How Adverse Employment Shocks Affect Marriage, Fertility, and Children's Living Circumstances. NBER Working Paper, 2015.

[iv] Raven Molloy et al. "Understanding Declining Fluidity in the US Labor Market." *Brookings Papers on Economic Activity*, forthcoming in 2016.

[v] Isabel Sawhill and Quentin Karpilow, *Raising the Minimum Wage and Redesigning the Earned Income Tax Credit*, Brookings Institution, 2014; Congressional Budget Office, *The Effects of a Minimum Wage Increase on Employment and Family Income*, 2014; Isabel Sawhill, *Paid Leave Will Be a Hot Issue in the 2016 Campaign*, Brookings Institution, 2015; Henry Aaron, *Five Years Old and Going on Ten: The Future of the Affordable Care Act*, Brookings Institution, 2015;and Harry J. Holzer, *Higher Education and Workforce Policy: Creating More Skilled Workers (and Jobs for Them to Fill)*, Brookings Institution, 2015.

[vi] Anne Roder and Mark Elliott. *Stimulating Opportunity: An Evaluation of ARRA-Funded Subsidized Employment Programs*. Economic Mobility Corporation, 2013.

QUESTIONS FOR THE RECORD FOR DR. MCAFEE SUBMITTED BY SENATOR AMY KLOBUCHAR

IMMIGRATION REFORM AND THE ECONOMY

Dr. McAfee, in your testimony, you stated that many of the world's most talented people want to come to the U.S. and build lives and careers, but our policies are often too restrictive.

Immigrants have been part of our nation's greatest achievements. Seventy-three of the Fortune 500 companies were founded by immigrants and even more were founded by immigrants or their children, including 3M, Best Buy, and Mosaic in Minnesota.

- How can comprehensive immigration reform benefit the U.S. economy?
- How can immigration reform help make the U.S. more competitive globally?

IMMIGRATION REFORM AND INNOVATION

Dr. McAfee, we know that immigration reform will also help spur innovation. Over 25 percent of all U.S. Nobel laureates were foreign-born. And as noted in this year's Economic Report of the President, one-quarter of all U.S. technology and engineering companies started between 1995 and 2005 were founded by immigrants.

- What policies should be implemented to make sure that the U.S. is attracting and retaining the world's talent?

INFRASTRUCTURE INVESTMENT

Dr. McAfee, in your testimony, you discussed the importance of sensible public spending on infrastructure and how fixing our crumbling roads and bridges would help increase our productivity growth.

Infrastructure from our ports, bridges and roads and water treatment facilities, and making sure that we have safer trains and efficient air travel is essential to the U.S. economy. Infrastructure serves as the foundation to support our country's economic global competitiveness and connects communities and people.

A well-maintained, efficient transportation system is essential to the future economic competitiveness of the U.S.

I was one of the first Senators to support the FAST Act which authorizes significant levels of investment of $306 billion in the nation's transportation infrastructure with Minnesota receiving more than $4 billion over the next five years. These funds will be used for our highways, bridges, rail and mass transit.

- Please discuss how investments in infrastructure support the middle class and keep the U.S. competitive.
- What other policies would you recommend to make sure that we have a 21st infrastructure and Internet?

QUESTIONS FOR THE RECORD FOR DR. HOLZER SUBMITTED BY SENATOR AMY KLOBUCHAR

IMMIGRATION REFORM AND INNOVATION

Dr. Holzer, we know that immigration reform will also help spur innovation. Over 25 percent of all U.S. Nobel laureates were foreign-born. And as noted in this year's Economic Report of the President, one-quarter of all U.S. technology and engineering companies started between 1995 and 2005 were founded by immigrants.

- What policies should be implemented to make sure that the U.S. is attracting and retaining the world's talent?

INFRASTRUCTURE INVESTMENT

Dr. Holzer, in your testimony, you discussed the importance of sensible public spending on infrastructure and how fixing our crumbling roads and bridges would help increase our productivity growth.

Infrastructure from our ports, bridges and roads and water treatment facilities, and making sure that we have safer trains and efficient air travel is essential to the U.S. economy. Infrastructure serves as the foundation to support our country's economic global competitiveness and connects communities and people.

A well-maintained, efficient transportation system is essential to the future economic competitiveness of the U.S.

I was one of the first Senators to support the FAST Act which authorizes significant levels of investment of $306 billion in the nation's transportation infrastructure with Minnesota receiving more than $4 billion over the next five years. These funds will be used for our highways, bridges, rail and mass transit.

- Please discuss how investments in infrastructure support the middle class and keep the U.S. competitive.
- What other policies would you recommend to make sure that we have a 21st century infrastructure and Internet?

1. What policies should be implemented to make sure that the U.S. is attracting and retaining the world's talent?

There is no question that attracting and retaining highly-educated foreign workers, particularly in STEM fields, contributes to innovation, job creation, and productivity/earnings growth in the U.S.[1]

To encourage more foreign-born STEM workers to enter and remain in the U.S., we should:

- Reform the H1B program—Current admissions of foreign STEM workers for temporary employment under the H1B program are capped at 65,000 per year, which is too low. I support raising the cap and the entry fee for employers, perhaps using a lottery process to determine both. The fees generated could be spent on increasing the education of native-born STEM workers in the U.S., particularly among minority populations. I would also make it easier for such workers to change employers in the U.S., though this might reduce the extent to which employers value the program; and I would make it easier for them to apply for permanent residence here.
- Adopt a Merit Point System for Annual Immigration—This idea was part of the Senate's Immigration Reform proposals in 2013. Instead of immigration being based primarily on family unification, adoption of a merit point system would make it easier for highly educated immigrants to obtain permanent residence status in the U.S.
- Exempt Highly Educated Foreign STEM Workers from Caps on Employment-Based Immigration—This idea was also contained in the reform proposals of the Senate bill.

2. Please discuss how investment in infrastructure support the middle class and keep the U.S. competitive.

Maintaining high productivity growth in the U.S. is a necessary condition for growth in our living standards, particularly for the middle class. A lengthy literature by economics researchers shows that infrastructure spending contributes to productivity growth in the U.S.[2] Given the declines in infrastructure spending in the U.S. in recent decades, as well as the decline in our productivity growth over the past decade (Baily and Bosworth, 2015), it seems like the economic returns to infrastructure spending right now would be very high.

An important additional positive effect of infrastructure spending in the short- to medium-term would be the creation of many more good-paying jobs for high school graduates in the U.S., particularly among men and in construction. Middle-skill and middle-wage employment has fallen in the U.S. in the past few decades, particularly for men without postsecondary education (Autor, 2010, 2015; Holzer, 2015). Construction employment has declined in the U.S. since the beginning of the housing crisis in 2006; it has recently recovered partially but not fully.[3] Expansion of infrastructure development would boost the availability of such employment. With many Baby Boomer construction workers soon retiring, such an expansion would require us to train a new generation of such workers, which would be of great benefit to young men who are currently leaving the U.S. labor force in droves because of their

---

[1] Hunt and Gauthier-Loiselle (2010) indicate that a 1-percentage point increase in the immigrant college graduates' population share raises patent production in the U.S. by 9–18 percent. Giovanni Peri et al. (2014) also show a 1-percentage point increase in the foreign STEM share of a city's employment increases the wages of native college graduates by 7–8 percent and of native non-college employees by 3–4 percent.

[2] See Berndt and Hansson (1991), Nadiri and Mamuneas (1994); and Holtz-Eakin and Lovely (1996).

[3] According to the U.S. Bureau of Labor Statistics, construction employment in the U.S. peaked at 7.7 million in 2006. It declined to 5.5 million in 2010 and has since recovered to only 6.6 million. Thus, only half of the construction jobs lost in the Great Recession have returned to date.

weak earnings prospects. Expansion of apprenticeship opportunities in the construction trades would be particularly helpful here.

3. What other policies would you recommend to make sure that we have a 21st century infrastructure and Internet?

While the issue of how to fund such infrastructure growth remains hotly debated, a recent report from the bipartisan Congressional Budget Office (2016) offers clues about how to make sure infrastructure spending is efficiently implemented. And, while Internet policy is generally outside of my realm of expertise, some recent reports from the Progressive Policy Institute on this issue seem persuasive.[4] Of course, ensuring that U.S. workers, at both the sub-BA and BA levels, maintain their computer skills is critical to ensuring that the benefits of the Internet are widely shared with U.S. workers.[5]

Autor, David. 2010. *The Polarization of Job Opportunities in the U.S. Labor Market: Implications for Employment and Earnings.* Hamilton Project, Brookings Institution, Washington DC.

Autor, David and Melanie Wasserman, 2015. *Wayward Sons: The Emerging Gender Gap in Labor Markets and Education.* Washington DC: Third Way.

Baily, Martin and Barry Bosworth. 2015. "Productivity Trends: Why is Growth So Slow?" Presentation, Brookings Institution, March 15.

Berndt, Ernst and Bengt Hansson. 1992. "Measuring the Contribution of Public Infrastructure Capital in Sweden." *Scandinavian Journal of Economics.*

Congressional Budget Office, 2016. *Approaches to Making Federal Highway Spending More Productive.* Washington, DC.

Holtz-Eakin, Douglas and Mary Lovely. "Scale Economies, Returns to Variety, and the Productivity of Public Infrastructure. *Regional Science and Urban Economics,* Vol. 26.

Holzer, Harry. 2015. *Job Market Polarization and U.S. Worker Skills: A Tale of Two Middles.* Brookings Institution, Washington, DC.

Hunt, Jennifer and Marjolaine Gauthier-Loiselle. 2010. "How Much Does Immigration Boost Innovation?" *American Economic Journal: Macroeconomics.* Vol. 2, No. 2.

Mandel, Michael. 2016. *Long-Term U.S. Productivity Growth and Mobile Broadband: The Road Ahead.* Progressive Policy Institute, Washington, DC.

Nadiri, M. Ishaq and Theofanis Mamuneas. 1994. "The Effects of Public Infrastructure and R&D Capital on the Cost Structure and Performance of U.S. Manufacturing Industries." *Review of Economics and Statistics.*

Nager, Adams and Robert Atkinson, 2016. The Case for Improving U.S. Computer Science Education. Information Technology and Innovation Foundation, Washington, DC.

Peri, Giovanni, Kevin Shih and Chad Sparber. 2014. "Foreign STEM Workers Boost Wages for U.S. Workers." National Bureau of Economic Research Working Paper.

QUESTIONS FOR THE RECORD FOR MR. ADAM KEIPER SUBMITTED BY SENATOR TOM COTTON

- There are some war fighting functions that machines can carry out more quickly and accurately than humans (such as targeting or engaging threats defensively). Other actions are not as easily automated (determining combatants or proportionality in combat). As technology advances, how should we create regulations about what decisions humans must make in warfare, and operations that machines should not do?
- As machines increasingly have a place on the battlefield, how do we grapple with the rates of PTSD among those operators removed from direct combat? How do robotics and automation increase or decrease the ethical complexity of warfare?

[Mr. Keiper's response was not received before the hearing was printed.]

---

[4] See Mandel (2016).
[5] See Atkinson (2016).

www.ingramcontent.com/pod-product-compliance
Lightning Source LLC
Chambersburg PA
CBHW080000230526
45470CB00008B/2808